THE CANALS OF NEW YORK STATE

(An American Canal Society Publication)

Some of our associates in the State of New York recently suggested that we consider publishing a book, in the style of our four BEST FROM AMERICAN CANALS, devoted exclusively to the canals of the Empire State, past and present.

On paging through all the issues of AMERICAN CANALS, since our first edition in 1972, we discovered that we had a surprising amount of editorial and pictorial material available – much of it contributed by the members of the American Canal Society.

Hence, this book!

Once again, we are indebted to these ACS members for the information published on the following pages, as well as to the Canal Society of New York State, from whose files we have extracted some of the information on those New York Canals not covered in our own periodical.

The articles published here are credited to the individuals who wrote them. It is my experience that all these devoted canal enthusiasts have done their research with great care, and their writings are as accurate as possible.

York, Pennsylvania
October, 1995

William H. Shank, P.E.
Publisher AMERICAN CANALS

COVER ILLUSTRATION

The artist who created our cover was William H. Bartlett, born near London, England in 1809. He came to America to sketch American landscape scenes in 1836. Among them was this familiar Lockport, New York, flight of locks. His engravings appeared in a book entitled "American Scenery" in 1838. The print used on our cover was more recently reproduced, in color, by the Mid-Continental Telephone Corporation, to whom we are indebted for making it available to us.

First Printing May 1991
Second Printing Oct. 1995

Published By

The American Canal and Transportation Center

American Canal & Transportation Center
333 Beach Drive
Annapolis, MD 21403

ISBN-O-933788-82-7

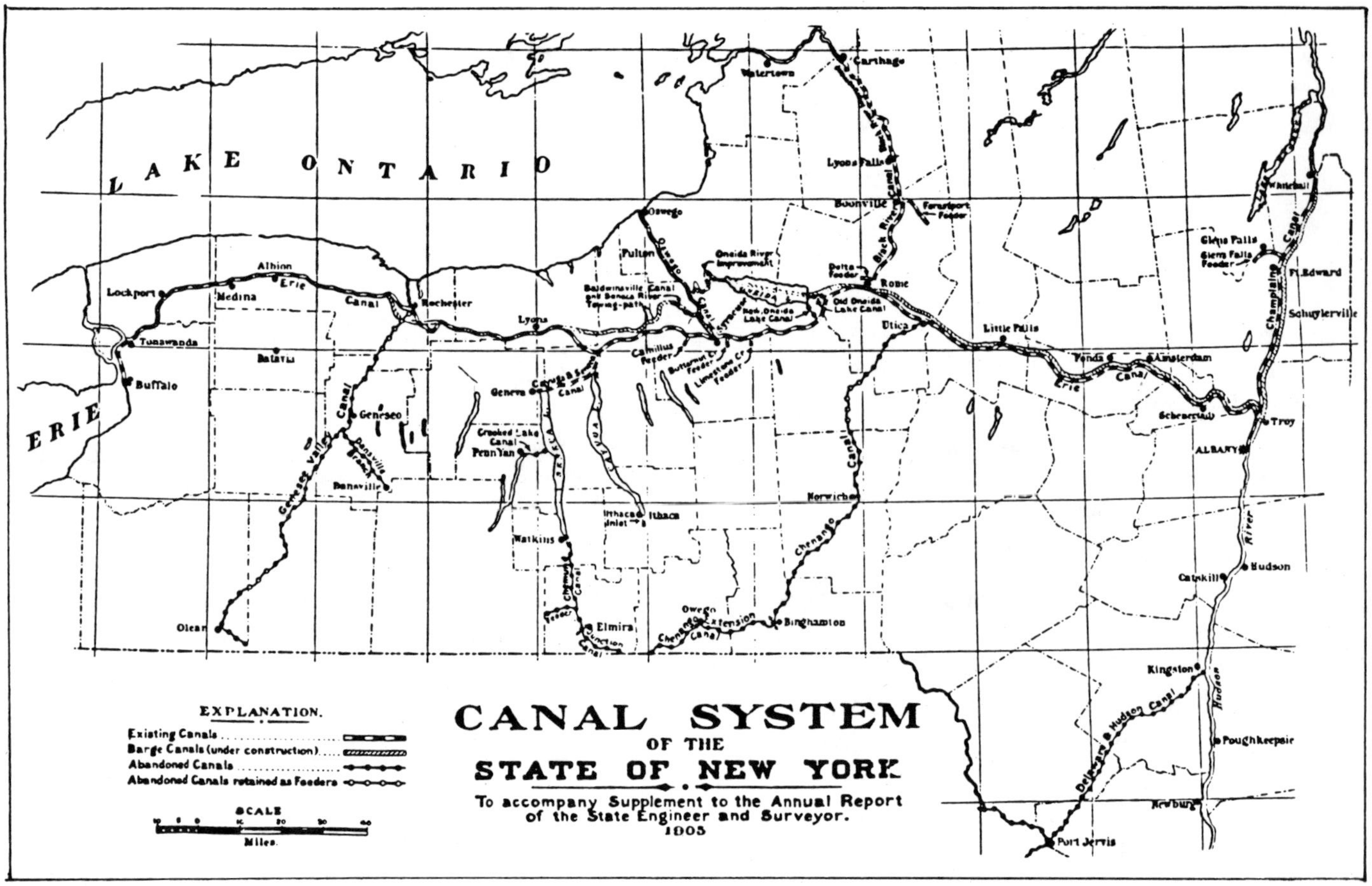

TABLE OF CONTENTS

ERIE CANAL – 170 YEARS OLD!

Horses pull a passenger-packet boat along a 1.5 mile restored section of the old Erie Canal near Erie Canal Village in Rome, New York. (Courtesy New York State Department of Commerce.)

This year New York State celebrates the 170th Anniversary of the opening of the Erie Canal (1825). This event created, for the first time, a direct navigable water connection between the Atlantic Ocean and the Great Lakes; turned the little town of New York into the nation's leading seaport; and touched off a flurry of canal building in the Northeastern United States unequaled anywhere else in the World since the British canal-building era fifty years earlier.

At the turn of the Nineteenth Century, construction of tow-path canals in America had languished. Except for extensive sales of lottery tickets for the Union Canal, and the Albert Gallatin Report of 1808, there was little to maintain public interest in canal building. Money, too, was lacking and the War of 1812 with England diverted attention for several years from internal improvements.

Only in New York State was interest in canals kept alive -- first by such interested parties as Elkanah Watson, who conducted a personal public relations campaign for a water connection between Albany and Lake Ontario; later by General Schuyler and William Weston, who in 1797 made an exploratory tour of upper New York; by James Hawley, a prominent New York Citizen who wrote a series of newspaper articles (1807) on the value to the city of a navigable waterway between the Hudson and Lake Erie; and -- finally -- the redoubtable DeWitt Clinton, mayor of New York City and later State Governor.

A delegation of two New York State legislators, Judge Forman and William Kirkpatrick, had approached President Thomas Jefferson in Washington, after the Gallatin Report had been made public, seeking some of the $20 million mentioned in the report. Jefferson, who had just signed a bill (1806) to get work started on the famous National Road was cold to the entire idea of the Erie Canal. "It is a splendid project" he said "and may be executed a century hence . . . . here is a canal of a few miles, projected by General Washington (The Potomack Canal) which has languished for many years because the small sum of $200,000 . . . cannot be obtained. And you talk of making a canal three hundred and fifty miles long through a wilderness! It is little short of madness to think about it!"

Geddes Runs Surveys

But there were many prominent citizens of New York who would not give up on the idea. About 1808, the New York legislature appropriated $600 for a survey, which Judge James Geddes, of Onondaga County, was asked to run. Geddes, a lawyer, with some experience in surveying, ran surveys from Albany to both Lake Ontario and Lake Erie, using the natural waterways as much as possible. When he made his report a year later, Geddes argued that if the Lake Ontario connection were used, much of the trade might be turned aside into the St. Lawrence River, whereas a direct connection with Lake Erie would assure through trade with the western states in the USA. The War of 1812, and the threat of further hostilities to the North, crystallized the determination of the Legislature to pursue an all-American route to Lake Erie.

There was considerable opposition to the Erie Canal, particularly by representatives from New York City, since it now appeared that the full expense of such a project would have to be borne by the State, with no help from the federal government. DeWitt Clinton, the man later to be called "the builder of the Erie Canal" had at first expressed little interest in the project. His friends won him over, and he soon became one of the most vigorous promoters of the Canal. At the close of the War with England, a group of responsible citizens, headed by DeWitt Clinton, met in New York City (1815) and signed a petition explaining the benefits of the proposed canal, which was circulated throughout the State, particularly to those counties through which the canal would pass. As a result the State Legislature, at its next session, received appeals from more than 100,000 of its constituents to get the canal started! They subsequently appropriated $20,000 for detailed surveys. In the meantime, DeWitt Clinton, who had served previously as Mayor of New York, was elected (1817) Governor of the State, putting him in the most favorable position to see the project through. He predicted that the Canal would be finished in ten years, and was re-elected Governor in 1821, as it became obvious that his ten-year prediction was conservative.

(Continued on Page Four)

De Witt Clinton, usually referred to as "The Builder of the Erie Canal". (From Alvin Harlow's "Old Towpaths")

(Continued from Page Three)

It was decided that not only would the Canal be built across northern New York State, but that a separate Canal would also be dug to connect the Tidewater Hudson (near Albany) with Lake Champlain, in anticipation of a connection into the St. Lawrence via the Richelieu River. For the final survey work, Benjamin Wright was assigned to the Erie, and James Geddes to the Champlain. While these two men had both had some surveying experience, neither of them could lay claim to the title of Engineer. Yet the two of them assumed the technical direction of the biggest engineering job yet attempted in America.

Canvass White

Canvass White, a young engineer with the Wright survey team, turned out to be the real engineering genius of the Erie Canal. He had studied surveying, mathematics, astronomy and other subjects at Fairfield Academy. His work in running levels west of Rome, N. Y., soon came to the attention of Benjamin Wright, and later, DeWitt Clinton himself, who saw rare possibilities in the young man. At Clinton's suggestion, Canvass White was sent to Europe in the fall of 1817 to inspect the canals of the Old World and to obtain some up-to-date surveying equipment. White walked 2000 miles along the canals of Great Britain, studying every feature. He returned the following year with copious notes and drawings, and new instruments.

In the meantime there was great indecision as to whether to build the locks of wood or stone. Proper cement for the stone could be procured only from Europe, and at great expense; wood would be perishable in a few years. Doubtfully, the decision was made to use stone, putting the blocks together with quicklime mortar, and using the expensive European cement only for "pointing." However White, within a few months after his return from abroad, discovered a deposit of stone near Chittenango, on the line of the canal, from which an excellent grade of hydraulic cement could be made!

The actual work of construction began (July 4, 1817) at the center section of the canal, where the going was easiest. By December of 1817, between 2000 and 3000 men were at work and 15 miles of canal had been completed. A number of Yankee innovations had been developed such as the plow and scraper, dumping wheelbarrows, sharp-edged shovel, etc. to speed up the manual work. An unusual tree-cutter, as well as a stump puller were invented on the Erie Canal. Difficulties were experienced in the Montezuma marshes near Syracuse, where the ground was so saturated with water that excavation was postponed till winter when the ground was frozen. In the summer of 1819, a thousand men died in the same area, of malaria. The lessons of the Panama Canal had still to be learned, and the importance of eliminating mosquitoes! In spite of many set-backs, and constant heckling by the opponents of the canal, the work continued with short sections of canal being opened to boat traffic as they were completed. One of the most difficult feats of the entire project was the climbing of the Niagara Encarpment, a solid-rock ridge at the west end of the route. Here Nathan Roberts pierced a solid mixture of limestone and flint, using black blasting powder to cut a flight of five double locks, each with a lift of twelve feet – thus producing the famed Lockport Locks. It took two years to do it, but when completed, every man on the job was proud of every lock that they had blasted out of nearly solid rock. This was the major obstacle to join the canal with the waters of Lake Erie, at Buffalo.

When the canal was opened on October 26, 1825, nearly two years earlier than planned, the populace of the entire State declared a holiday, while De Witt Clinton and a triumphal group of dignitaries officially opened the canal with a flotilla, starting in Buffalo and winding up in New York City, preceded by cannon-fire along the entire route, with a "wedding of the waters" ceremony at the New York City end. De Witt Clinton personally dumped a container of Lake Erie water into New York Harbor, symbolic of the all-water connection to the Great Lakes. It was a great day in the history of the State of New York, marking the completion of one of the most ambitious engineering achievements of mankind to that date.

Champlain Canal

Almost overlooked in the excitement of the last few months of the building of the Erie, was the opening in 1823 of the Champlain Canal. The architect of this project was James Geddes, assisted by a French Engineer named Marc Isambard Brunel, who had done the original survey work between the Hudson River and Lake Champlain. Construction on the Champlain Canal began in 1818 and it extended sixty-six miles upstream from two entry points: one, a series of locks off the Hudson River in the village of Waterford; and the other, at the Cohoes Junction with the Erie Canal, via the Waterford southern tier. Northern terminus of the canal was Whitehall, on a navigable arm of Lake Champlain. The Lake extends north across the Canadian border, where it becomes the source for the Richlieu River, running north to the St. Lawrence.

Erie Canal Aqueduct at Rochester, New York, about 1900. (Courtesy William Etchberger, Lebanon, Pa.)

After normal relations had been re-established with the Canadians at the close of the War of 1812, the Legislature of Lower Canada passed a bill (1818) granting authority for a canal to by-pass the Chambly Rapids on the Richlieu. A lock and dam had previously been built at St. Ours, to permit slack-water navigation along the Richlieu to the Chambly area. Numerous delays and financial difficulties beset the Chambly Canal, which was initially opened in 1843, but due to poor construction, had to be rebuilt. It was finally opened to through traffic in 1858, 12-miles in length, with 9 locks.

Thus in 1858, a direct water connection was completed between the Port of New york and the Ports of Montreal and Quebec on the tidewater section of the St. Lawrence River – a major achievement for both countries, and the beginning of expanded trade between them. Needless to say, the economy of the Champlain Valley in northern New York State and Vermont was favorably affected by this new, inland route to upper Canada.

In the early Nineteenth Century the financing of the Erie Canal turned "Wall Street" into the economic center of America. (Courtesy United Airlines Magazine.)

"Wall Street"

The financing of the Erie and Champlain Canals was a colossal undertaking for the State of New York. Refused any help by the federal government, they decided to "go it alone". Banking and investment firms in lower Manhattan were set up whose express purpose was to sell stock and collect money for the financing of the State's canal-building program. The area later became known simply as "Wall Street". For a description of this new financial development in the USA, we quote Thomas Kiernan, author of "The Road to Colossus":

"Not until after the War of 1812 did the Wall Street symbol take shape. Until then, Philadelphia and Boston were the country's main financial centers – Philadelphia because that was where the first federal bank was established, Boston because that city remained the nation's busiest port and chief source of private investment finance and banking. It was only when the state of New York set out in 1817 to raise funds for the construction of the Erie Canal that New York City – Wall Street – began to rise to prominence as a national financial hub. Investment syndicates and banks in Boston and Philadelphia had followed a tradition of aristocratic exclusivity in their private financing operations, limiting participation in important investment schemes to select members of the moneyed classes.

"The financing of the Erie Canal was by far the largest fund-raising endeavor ever attempted. Because the canal was to be a uniquely New York enterprise, designed to improve the economic fortunes of the city and state at the likely expense of Massachusetts and Pennsylvania, it was roundly resented in those regions and failed to attract significant funds from exclusivist Boston and Philadelphia sources. Thus, almost by default, the financing of the canal became the first instance in which ordinary citizens and organizations invested their money in the hope of future profit.

"The fact that the state guaranteed every individual investment in the canal project sweetened the lure, and the response was overwhelming. Because New York City was the chief locus of the state's private banking, trading, and legal network, most of the financing was conducted from there. Private banks and one-man investment firms were created to broker the canal funding, and storefront offices were established along Wall Street and its tributary lanes to handle the complex handwritten paperwork connected with it. Soon the district was saturated with a dense mix of private banking offices, money exchanges, canal-business firms, and the like.

Canal A Success!

"Wall Street might have remained just another busy small-city Main Street had it not been for the immediate success of the Erie Canal and the resultant canal fever it generated in the rest of the country. The Erie recovered its construction costs out of revenues within a few years and began to pay handsome dividends to its investors. Since Wall Street had demonstrated such expertise in canal financing, canal builders from other regions called upon New York money and brokerage firms to handle their financing.

"The success of the Erie also made New York the preeminent commercial city in the nation generally, and more and more foreign investment money gravitated to Wall Street. By 1830, New York's stock exchange, dealing primarily in canal securities but also promoting shares in new industrial companies formed to exploit the transport benefits provided by the Erie, far outstripped the Boston and Philadelphia exchanges."*

NY Canals Today

Today, even though there is occasional talk of abandonment, which has spawned the formation of such organizations as BARGE (Bi-Recreational Association for the Restoration of the Great Erie) the Inland Waterway System of New York State appears to be alive and well.

The State of New York (Department of Transportation) boasts the only waterway system in the country which, (with the exception of one C.E. lock on the Hudson River), does not depend upon the Corps of Engineers for its operation and maintenance. The entire Waterway consists of the Erie Canal, the Champlain Canal, the Oswego Canal, and the Cayuga and Seneca Canals.

It's commercial transport consists primarily of oil barges (built to fit the locks) and a few other vessels, such as Capt. Olaf Kaldefoss "M.V. Day Peckenbaugh," carrying cement. There are also sizable passenger-carrying boats, such as Peter Wiles' "Emita II" and the American-Canadian Lines' "New Shoreham II"

(Concluded on Page Six)

**For full information on the development of the NYC financial district, read "The Road to Colossus" published by William Morrow & Co., Inc., 1985.*

Erie Canal – 170 Years Old!

(Concluded from Page Five)

which make regularly scheduled trips over most of the New York Waterways.

In recent years, the Erie Canal has experienced a huge surge in its use for recreational purposes. In 1984 alone, nearly 100,000 pleasure boats were locked through the 36 canal locks and tour boats provided thousands of passengers with a new appreciation of New York's beautiful landscapes from the unique perspective of a canal cruise. Boating and "locking through" has been free since 1883, when tolls on the canal were dropped.

As of this writing, the following events are planned in connection with the 160th celebration of the opening of the Erie Canal:

On June 10, a one half-hour video documentary, "Carry On, Oh Erie," (The Extraordinary Course of the Erie Canal) will be premiered in Albany. During the summer months, it will be seen on PBS stations throughout the nation. With Burl Ives narrating, the documentary, produced by Van der Broeck Associates Inc., will highlight the Erie Canal's historical significance to New York State and its expanding importance as a recreational waterway and energy resource.

June events include a Rensselaer County Riverfront Festival in Troy on June 1-2, the Empire State Regatta in Albany on June 8-9 and a Village Days Festival in Fairport on June 13-14.

Bill Shank, and party, stopped in recently at the Erie Canal Museum in Syracuse, New York. Extensive construction work is being done in the north section of the Museum, where the old weigh lock is being unearthed and a full-scale replica of an old Erie Canal Boat is being built. The entire project is expected to be completed by the end of 1985. Bill is shown here, with Todd Weseloh (left) – a valued member of the museum staff – in front of one of the many museum displays.

On July 4, the re-enactment of the Erie Canal ground breaking will be staged at the Erie Canal Village in Rome. On July 13, at Van Buren, Onondaga County, an "Anything That Floats Day" observance will be held. A Canal Fest '85 is scheduled at Tonawanda - North Tonawanda on July 21-28.

A Hudson River Festival will be held in Coxsackie, Greene County, on Aug. 3, a Village Sesquicentennial Celebration is planned in Clyde, Wayne County, on Aug. 12-15, and a "Canal Day" observance will occur at the Schoharie Crossing State Park in Fort Hunter, Montgomery County, on Aug. 18.

In September, Canal Days will be staged in Palmyra on Sept. 14-15 and a Towpath Rally will take place at Rochester on Sept. 21.

The 160th Anniversary of the Opening of the Erie Canal will be observed in New York City, Albany and Buffalo on Oct. 26.

Canal Museums

The flavor of the early canal days in New York State, and the role canals played in the nation's social, cultural and economic growth, are recaptured in several museums and reconstructed canal towns in the State.

Chief among these is the Erie Canal Museum located in the Weighlock Building in Syracuse, the last remaining structure of its kind to survive from New York's 19th Century canal system. Constructed in 1849-50, it housed the scales used for determining canal tolls.

The museum's orientation center features lock and boat models, tools, original engineering drawings, prints, paintings and photographs. Special canal-oriented exhibits are changed periodically. A $400,000 capital improvement program is to be completed by the end of the year. New attractions will include a replica of a full size mid-19th Century canal boat, hands-on exhibits and a slide orientation program.

The Erie Canal Village at Rome, near the spot where the first shovelful of dirt was turned for the Erie Canal, is a circa-1840 canal village reconstructed in a rustic setting. Its principal attraction is a mule-drawn passenger packet ride along a 1.5-mile restored section of the original canal. An 1890-era narrow gauge steam locomotive and train also operates on one bank of the canal.

The Canastota Canal Town Museum, in the Village of Canastota, also offers a glimpse of life in the "old canal days." Open to the public year-round, from 9 a.m. to 4 p.m. Monday through Friday, and weekends on request, with no admission. A mid-19th Century building also has been restored to its original condition and includes furnishings, photographs and models from the Erie Canal era.

Erie Canal brochures are available from the State Department of Transportation, Office of Communications, 5 Gov. Harriman State Campus, Albany, N.Y. 12232.

ERIE CANAL BICYCLE TOUR

Classic Bicycle Tours of Clarkson, New York is offering a bicycle tour along the Erie Canal August 6 - 13, 1989. The tour will begin in Rome, New York at the exact spot where the first shovel full of earth was turned by Governor DeWitt Clinton on July 4, 1817.

After a visit to the Erie Canal Village and a ride on a horsedrawn canalboat, the bicyclists will follow the Old Erie down to Syracuse. From Syracuse, the tour will continue westward and finish in Lockport, New York with a dinner cruise through the famous Lockport locks.

The tour is billed as a family trip with wonderful educational attractions, level terrain, low daily mileage (30 miles or less) and over half the trip is on canal paths separate from road traffic.

Cost of the tour is $629 (exclusive of travel to Rome, NY). This cost includes accommodations, breakfasts, welcoming and farewell dinners, entrance fees to museums, boat rides on the canal, a support vehicle that carries all luggage, and experienced guides and bicycle mechanics.

For more information write or call (toll free): Classic Bicycle Tours, P.O. Box 668, Clarkson, New York 14430; 1-800-777-8090 or (716)-637-5970.

Contrary to the usual opinion, the Erie Canal was not the first Mohawk valley canalization project. Archaeologists are now beginning to excavate the remains of various short canals built 30 years earlier to circumvent obstacles to navigation on the Mohawk River, according to an article by David Germain published in the Daily Messenger of Canandaigua, N.Y. on October 11, 1990. Philip Lord of the State Museum in Albany is in charge of the work. (Contributed by Walter Meseck)

ERIE CANAL MUSEUM PLANS EXPANSION OF DISPLAY FACILITY

A one million dollar expansion of the Erie Canal Museum at Syracuse is being funded by the State of New York in cooperation with the County and the City of Syracuse. The new two story addition will be about one third the size of the original Weighlock building that presently houses the museum.

This added facility will provide space for a new entrance, a theater, elevator access, heating and cooling and an expanded gift shop. Ground breaking is planned for the spring of 1990 with a grand opening scheduled for December of that year.

A major special exhibit planned for the museum, A Novel Look at the Erie Canal, is scheduled to open on June 6, 1989. It will feature a fictional look at the history of the canal that is not possible in nonfictional literature.

ERIE CANAL IS HAVING TROUBLE COPING

The fabled Erie Canal, made famous to schoolchildren through the song about the mule named Sal, is having trouble coping with the 1970's. The state Barge Canal, the successor to the engineering marvel of the early 19th Century, now faces an uncertain future.

With its locks too small for modern barges, and competition from other modes of transportation, the canal needs a major renovation, or will serve only pleasure craft. Some even predict it could be shut down within a decade if not improved.

The condition is a far cry from the time the canal system was the major link between the Atlantic Ocean and the nation's interior. Cities like Syracuse, Rochester and Utica grew along the canal which also helped make New York City the economic center it is today.

When the original four-foot deep canal opened in 1825 it was hailed as the greatest public works project of a free society. It was the longest canal in the world.

The canal lowered transportation costs. It also tied the Midwest to the Northeast, rather than the South and the Mississippi River, a significant factor in the outcome of the Civil War, according to historians.

In its heyday, traffic was so heavy the canal not only paid for itself, but also financed state government, until it was made toll-free in 1883.

The canal gave New York City a big advantage over Boston, Philadelphia and other port cities, drawing business to New York City with its access to the Midwest. But the glory of the canal system has tarnished in the face of competition from highways, railroads, airports and the big freighters of the St. Lawrence Seaway.

Now only about 40 barges regularly carry cargo over the 527-mile canal system. After more than 150 years in operation the historic canal faces a bleak, and possibly limited, future.

Although the state still spends about $10 million a year to keep the canal system operating, a state Senate task force recently warned the barge canal may be closed within a decade if it is not modernized. A U.S. Senate subcommittee scheduled a hearing on the canal's future in Washington in August.

The fabulous double flight of ten locks on the old Erie Canal at Lockport, New York, has been partially replaced (1915-18) by two, fifty-foot lift locks of the Erie Barge Canal (at left). One of the old "flights" has been preserved (at right) partly for its historic value, and partly to serve as a spillway to bring Lake Erie water to the upper reaches of the Barge Canal. The old lock gates have been replaced with vertical weirs. (Photo by Tom Hahn.)

Interior of the guillotine type lift lock, highest lock on the eastern half of the Barge Canal, at Little Falls, N.Y. (Photo by Bill Shank.)

The 57 locks on the 14-foot deep canal are too small to allow the big barges that ply the waters of the New York Harbor and Great Lakes to use the up-state canal. Some of the locks are over 60 years old. The average barge using the canal is two decades old and the newest has been in operation five years, the state senate task force reported.

A tug boat on the Ohio River can move 20,000 tons of cargo. A tug on the Mississippi River can move 50,000 tons. But a tug on the New York barge canal can handle only about 3,000 tons. Those physical limitations reduce the efficiency and economy of the barge canal, making it harder to compete with other forms of transportation.

The current canal, 14 feet deep and 140 feet wide, was built in 1915. It uses some sections of the old Erie Canal.

The expansion was done after the possibility of enlarging the canal for shipping was dismissed. Instead, the St. Lawrence River to the north was eventually dredged to accommodate ocean-going freighters, replacing the canal as the major water link between the Atlantic and the Great Lakes.

Traffic on the canal first declined in 1883 when the Erie barges began losing cargo to railroads. In 1951 the modern canal carried five million tons of cargo, its biggest year ever. But traffic on the canal has declined ever since. Last year, less than two million tons moved.

The canal remains a major transportation system for petroleum, especially industrial fuel oils. More than 94 percent of all cargo carried over the canal last year was petroleum. Molasses, cement, machinery and chemicals were also carried over the canal.

The canal itself has shrunk by almost 300 miles with branches being abandoned that once linked it to the cities of Port Jervis, Binghamton, Olean and Elmira, as well as a northern branch to Carthage.

The "Emita II", only commercial passenger vessel of modern times to cruise the entire length of the Erie Barge Canal. Modifications to the super-structure of the vessel this past summer made it possible for the "Emita II" to pass under the extremely low bridges of the Canal west of Syracuse. Captain Peter Wiles can be seen at the left. (Photo by Bill Shank.)

The canal's best hopes for a revival appear linked to the energy crisis. Rising fuel costs make the cheaper water transportation of the canal increasingly attractive. A canal barge can carry as much petroleum as 79 tanker trucks, with less pollution, but only at speeds up to 10 miles an hour.

The renewed emphasis on coal as an energy source has also raised hopes the canal may become a major conduit for western coal into the Northeast. However, coal has not moved on the canal in 16 years.

Although the canal is the only one in the United States not under federal control, dollars from Washington may be its best hope for modernization.

The state constitution prevents New York from giving up the canal, but there has been growing discussion of the need for federal help in updating the waterway.

(Submitted by ACS Member Daniel Mondell from the Batavia, N.Y. Daily News)

An Early Trip On The Erie Canal

by Ernest H. Schell

An old print, looking down one of the two flights of locks on the Erie Canal at Lockport, New York. (Courtesy George Wills)

The greatest monument to American enterprise in the early nineteenth century was the **Erie Canal,** the longest canal in the world in its day, a 364-mile giant over ten times as long as any canal previously built in the United States. It was a stupendous engineering feat undertaken by ingenious and dedicated amateurs working scientifically and systematically on a project that many people had considered a certain failure.

A host of visionaries had for decades promoted schemes that would link the Hudson River with the Great Lakes, but the bankruptcy of the Western Inland Lock and Navigation Company in 1808, after a disappointing 17-year history of marginally useful ventures, discouraged many of the state's leading merchants and speculators from investing in canal stock. While businessmen in New York City saw the value of having their town serve as the gateway to the West, they did not believe that a successful canal could be completed in their lifetime. Even DeWitt Clinton, the forty-one year old mayor of the city, when he was appointed to the New York State Canal Commission in 1810, was not enthusiastic about the prospects of the project. He had long recognized the value of a link with the West, however, and after the War of 1812 became the canal's leading promoter.

When construction of the Erie Canal was finally authorized in April, 1817, Clinton and the other Commissioners knew that to win over support from a wary public, and particularly from the financial community, a modest but successful beginning was preferable to failure on a spectacular scale. Accordingly, the canal was divided into three parts, with work on the middle section from the Seneca River to Utica to be completed by 1820. Under the direction of Judge Benjamin Wright, often considered the "father of American engineering," work proceeded on schedule, with Wright supervising the application of new equipment and innovative techniques in land-clearing, blasting, and construction.

Though foul weather damaged part of the uncompleted canal in the spring of 1819, and malaria and typhoid took their toll, all doubts about the project were dispelled by October, when a portion of the middle section from Rome to Utica was opened for traffic. On the 22nd, Clinton, now the governor of the state, embarked at Rome on a 60-foot long canalboat named **Chief Engineer of Rome,** in honor of Wright, leading a company of dignitaries who set off on the trip to Utica amid the pealing of bells, the salute of cannon, and the enthusiastic cheers of thousands who lined the canal banks to witness the great event. Investors were suddenly bullish about "Clinton's Ditch," vindicating the Commissioners' decision to bank on early signs of progress in winning the support necessary to complete the canal.

When the finished canal was opened with the famous "Great Pageant" in October, 1825, the attention of the entire country was focused on the event. By then, though, canal mania already had the country in its grip. Investment in the Erie had not only proven sound, but had been encouraged by the continued care with which the Commissioners had continued to promote their efforts. Had it not been for these early and elaborate celebrations, the "Great Pageant" itself might never have taken place

April 21, 1820, opening day for the entire middle section, for example, was as dramatically staged as the fanfare that had accompanied the opening of the Rome-to-Utica leg the previous October. The **Lion of the West,** a 76-foot boat with a 4-foot beam, left Rochester early in the morning with its contingent of the illustrious, laid over at Lyons that evening, and arrived in Utica the following day, having made the trip with an average speed of five miles an hour. Since wash from the boat had caused part of the bank to crumble along with way, a speed limit of four miles an hour was imposed to try to solve the problem. Otherwise, everything went as well as the Commissioners had prayed it would. On the 4th of July, an international exhibition at Syracuse commemorated the completion of what had already proven to be a resoundingly successful venture. The opening of the first commercially significant portion of the Erie Canal had become the most talked about event of the day.

The middle section was immediately put into use carrying freight and passengers. A ride on the canal, in fact, was considered high adventure for travellers, who could return home boasting of their experience. One of the passengers that first summer was a Philadelphia merchant, Thomas Pym Cope, on his way from Philadelphia to see Niagra Falls. Cope, civic leader who was later to become interested in several of Pennsylvania's internal improvements, was a meticulous observer of the world who for many years kept a diary of his daily rounds. Two passages from this journal, recently edited by Eliza Cope Harrison **(Philadelphia Merchant: The Diary of Thomas P. Cope, 1800-1851** [South Bend, Indiana: Gateway Editions, Ltd., 1978, $19.95]), recount the trip he took on the newly-opened middle division.

For some reason, Cope's trip terminated at Montezuma, near the Seneca River, which was in fact the officially designated western end of the middle division, although the canal had been open earlier in the month as far west as Rochester, and a Utica newspaper announced that on July 25, three days before Cope's journey, packet service was to have begun between Utica and Rochester. Unlike the inaugural voyage of the **Lion of the West,** Cope's canal sojourn continued through the night, a practice that became standard procedure, but one that could be dangerous, as Cope noted. There were other dangers, such as falling trees in stormy weather, that Cope noted as well. His account, repro-

(Concluded on Page Nine)

An aqueduct on the old Erie Canal, circa 1865-Hornung. (Early American Advertising Art)

An Early Trip On The Erie Canal (Cont.)

Sketch of the entrance lock to the old Erie Canal at Troy, New York, circa 1870. (Courtesy Canal Society of New York State.)

(Concluded from Page Eight)

duced in full with the permission of the publisher (Regnery/Gateway, Inc., South Bend, Indiana) offers a detailed and fascinating glimpse of the very early and exciting days on the Erie Canal.

"July 28, 1820: At 9 this morning we went on board the passenger boat **Montezuma,** Captn. Brown, on the great canal finished from Utica [where Cope's journey began] to Montezuma, distance of 96 mi. This boat is 75 feet in length, 12 beam and draws 20 inches of water. Captn. B. has $18 per mo. and says that for weeks he has not had 3 hours regular sleep at one time. This is hard duty for such wages. The boy who drives the horses has $10 per month. Today he is to drive 50 miles and must return tomorrow.

"Our boat is towed by tow horses attached to a line 110 feet long. Chains are not permitted, on account of their tearing away at the banks. There is but one tow path & when two boats meet, the one going to market stops & dropping her line, the other passes over it. Passenger boats pay 5 cents a mile, country produce one cent per 100 pds., and dry good 2 cents. The bridges over the canal, of which we are told there are 70 between Utica and Montezuma, are so contracted as to leave barely room for our boat, the abutments projecting too far into the canal.

"Captain B. is sick with the fever and ague and says all the lads in their employ on the canal – the one who now drives our horses excepted – have the same disorder. It is an unpleasant omen.

"One of our company has amused himself with hooking bullfrogs, which are numerous and display themselves in great stile on the banks. We have had a mess of them for dinner and deem the flesh not inferior to that of a chicken. The banks of the canal are falling in some places, owing to their having been made too steep. This error, however, they appear to have corrected.

"July 29: We made no stop at Manlius, which we passed late at night, just previous to which the Boy, getting asleep, fell from the horse and luckily was not much injured, but it was so dark that the accident was perceived only from the fright of the horses. There had been a heavy storm of wind and rain, accompanied by thunder. the **Oneida Chief,** another passage boat, bound to Utica, passed us some hours afterwards and reported that she was near being destroyed by the falling trees – indeed, the danger was considerable. About daylight this morning, we arrived at the first locks, 3 in number; they descend together 27 ft. to the west. Halted for half an hour at Syracuse, 61 miles from Utica – a village of 20 houses on the north side of the Canal.

"We were much too crowded last night. Captain B. says he can accommodate 10 or 12 passengers with comfort and does not wish more, whereas there are 21 of us. The women occupied all the berths and the men spread themselves on settees and on the floor and were said to have fared better than their fair companions, who complain of having been annoyed by vermin. Alfred [Cope's son], who slept on the floor, got up with one arm much swollen and inflamed, apparently from the bite of some insect.

Sketch of an Erie Canal Boat cabin interior, circa 1860. (Courtesy Canal Society of New York State).

"We reached Montezuma before twilight, 96 miles from Utica and have been 33 hours in coming that distance and therefore travelled at a rate less than 3 miles an hour. We had 4 meals on board and for these and passage paid $4 each. It has been a listless, dull voyage – the novelty alone rendering it tolerable. None of our company probably regret having taken it, but none, I believe, would be willing to renew the trip."

(Mr. Schell, a freelance writer and business historian, is a doctoral candidate in history at Temple University.)

ERIE CANAL ILLUSTRATIONS

Emily Madden, ACS, (canal researching), suggests that we try to include more information on the illustrations which we use in our articles. As a starter, she submits the following information on the four illustrations which were used in "An Early Trip on the Erie Canal" in the February issue of **American Canals.**

1. Old print of Lockport. The artist was William H. Bartlett, 1809-1854, an Englishman. The print dates from about 1828 and was published along with many of Bartlett's other prints in the two-volume work **American Scenery** by Nathanial P. Willis. This print was captioned "Lockport, Erie Canal" and is probably the best known of all the early prints at Lockport.

2. Aqueduct on the Erie Canal. The original artist is unknown, though there are surmises. The picture has been reprinted often; occasionally it has been captioned "Miami and Erie Canal." This print was reprinted in the **Centennial History of Rochester, New York,** V. II, 1932, being Vo. XI of the **Rochester Historical Society Publications** 1932, p. 306. It carried the following caption: Drawn before the completion of the structure, and printed by Everard Peck Rochester. Only seven of the eleven arches are shown. The iron railing, finally erected on the north side of the towpath does not appear; the Johnson-Seymour mill-canal on the east side of the river is not clearly traced. The point of view is on the west river bank, north of the structure, looking southeast."

This aqueduct was replaced in 1842 by the wonderful stone structure which is still in existence now carrying a four-lane street in the center of Rochester on top of it.

3. The two sketches on page 23. These were done by the famous artist Edwin A. Abbey (though he was not famous when the sketches were made in 1873.) He did the murals in the Boston Public Library, among others. He was born in 1854. These sketches appeared in an article by William H. Rideling, "The Waterways of New York" in **Harper's New Monthly Magazine** for December 1873.

(Editor's Note: Many articles submitted for publication in "AMERICAN CANALS" come to us without illustrations. We do the best we can to find suitable sketches or photos in our own files. Such was the case with "An Early Trip on the Erie Canal." We ask all contributors to supply us with good illustrations and complete caption material to accompany their text, whenever possible.)

Cruising New York State's Inland Waterways

The Emita II enters Lock Number "C-3" on the Champlain Canal, headed downstream for Albany. The sign-boards at the Champlain locks clearly indicate the Number and location of each lock, the water elevation above and below, and the mileage to the next lock.

by Bill Shank

The State of New York is perhaps the only place in the Northeast where inland commercial waterway traffic still exists over nearly the same routes followed by the historic canal system of 150 years ago. In the early 1800's the State asked for federal help in building its fabled "Erie Canal" – didn't get it – and went ahead on its own. It has been going it alone ever since! The State of New York is one of the few places in the United States where the Corps of Engineers (except for one, lone, tidewater lock at Troy, N.Y.) is *not* in control of the waterways. The rest is maintained and operated by the Waterways Maintenance Subdivision of the State Department of Transportation at Albany. It is known officially as the "New York State Barge Canal System". It includes the Erie Canal, from Troy to Towanda; the Champlain Canal, from Troy to Whitehall; the Cayuga and Seneca Canal, from Montezuma to Geneva; and the Oswego Canal, from Three Rivers to Oswego.

The entire 527-mile system is still open to commercial traffic, now consisting primarily of industrial fuel-oil barges, built to fit the locks (with inches to spare). In recent years, however, passenger traffic had been limited to small pleasure boats and a few tour boats, running for short distances out of major cities along the waterways.

About five years ago, the Mid-Lakes Navigation Company, Ltd, with headquarters in Skaneateles, N.Y., set about to offer charter "Packetboat Service" on various sections of the Barge Canal System. They began with scheduled three-day trips between Syracuse and Albany on the "Emita II," a former ferry boat. In 1978 they decided to extend their coverage to the western section of the Erie Canal, from Syracuse to Lockport. To do this, they had to make drastic alterations to the "Emita II" to pass under the extremely low bridges in the Rochester area, cutting off much of the superstructure, including a portion of the pilot house.

A year ago, they extended their chartered trips to include the Champlain Canal (3-day trip), as well as the Oswego Canal (1-day) and the Cayuga-Seneca Canal (2 days).

Having just returned from my third trip on the Emita II, along the Champlain Canal, I highly recommend any of these trips to canal buffs (and others) who wish to enjoy the beauty of up-state New York from the deck of a 50-passenger packet boat – not to mention the experience of "locking through" countless of the efficiently operated locks of the New York State Barge Canal System.

Don't expect to "dress for dinner" on board the Emita II; the atmosphere and relationship between Captain Peter Wiles, his crew and passengers is most informal. All meals are served, either on board, or in the shade of a convenient tree on shore. At night, the crew moves your baggage to a motel, or hotel, near the most convenient docking point. From time to time the Captain makes a "P-A" commentary on points of interest on shore, and is quite approachable to answer any questions, about almost anything, historical or current. He is a most knowledgeable and interesting personality. After twenty-four hours with passengers and crew on the Emita II, you begin to feel like "family" and can appreciate the close relationships which must have developed on the canal boats of 100 years ago!

For further information on the Emita II and its sailing schedules, write Mid-Lakes Navigation Company, Ltd., P.O. Box 61, Skaneatlles, N.Y. 13152.

"NARROW BOATS" ON THE ERIE CANAL?

The "Owasco", one of a fleet of Lockmaster canal boats, designed and built by Mid-Lakes Navigation of Skaneateles, New York. The boats are available for hire on the New York State Canals, and are similar in appearance and design to the famous English "Narrowboats".

NEW VISITORS' CENTER

On Tuesday, September 27, formally invited dignitaries and guests attended dedication ceremonies for the new Visitors' Center Building and its parking area and access road. The facilities were built by the Parks Department construction staff in its first phase of new development in the D&H Canal Park, as announced by Graham M. Skea, Commissioner of the Orange County (NY) Department of Parks, Recreation and Conservation.

The new handsome structure is close to the buildings that house the Neversink Valley Area Museum, in Cuddebackville, NY.

Mr. James F. Smith, Chairman and Chief Executive Officer of Orange and Rockland Utilities, Inc., presented to Orange County a deed to a nine-mile stretch of the D&H Canal towpath. This segment, which runs from Port Jervis to Westbrookville, is planned for hiking and bicycling uses, and will connect with State and Sullivan County property to create a "Greenway." Ultimately, this outdoor facility may reach all the way to Kingston.

A highlight in the new center will be the exhibit of a model of Roebling's Neversink Aqueduct, which has been built to scale by Museum member Jim Bennett. In the future, the center will be available for public use by permit.

Canal Book Reprint

The University of Kentucky Press has just issued a new edition of *ERIE WATER WEST-A History of the Erie Canal,* by Ronald E. Shaw. First printed in 1966, the new book contains 449 pages and an index, plus a few illustrations, bound at the book's center. An "Essay on Bibliography" is included. Clothbound price — $30; Paperback $15. Write The University of Kentucky Press, Lexington, Kentucky 40506.

NORTHEASTERN INLAND WATERWAYS TOUR

Bill and Ruth Shank pose for their photo on the deck of the New Shoreham II.

At the urging of several of our friends who have done it, including the Charles Hadfields of England, Ruth and I finally took the Saguenay River Cruise on the New Shoreham II, out of Warren, Rhode Island. We had found out, in the Spring of 1980, that it pays to "sign up" nearly a year in advance. (The Saguenay Tour is an extremely popular one.) We met George and Jeanne Irwin (from Camp Hill, Pa.) at Warren (just southeast of Providence) on June 28th and returned to Warren on July 10th, after having traveled approximately 1200 miles by water and 350 miles by land.

The New Shoreham II accommodates 72 passengers, and crew, and is a modern and well-equipped vessel with air-conditioned living and dining facilities, as well as a large and comfortable lounge. The crew did everything possible to make our cruise an enjoyable one. The young Chef served us an excellent variety of (not-too-heavy) meals, with lots of fruits and vegetables. (coffee and tea were available 24 hours a day!)

An array of draw-bridges greeted us at Lock #1 on the Oswego canal at Phoenix, N.Y. View into the lock chamber as we departed, downstream. Lock Lift: 10.2 ft.

Our cruise included a night tour of Long Island Sound, an early-morning look at Manhattan Island; a day-long trip on the Hudson River; passage through the 23 locks on the eastern half of the Erie Canal; the seven locks of the Oswego Canal; and the seven locks of the St. Lawrence River; not to mention visits to Waterford and Oswego, New York; Alexandria Bay (in the Thousand Island section); Montreal, Quebec, and the beautiful Saguenay River where the tides run 22 feet in height and the White Whales play at its junction with the St. Lawrence! For a canal and bridge buff (like me) it was a highly enjoyable and educational experience. I made literally hundreds of photos, in black and white and color, of the many interesting bridges crossing the Hudson and St. Lawrence, and of course interior shots of most of the canal locks. Some of the best of my photos are reproduced here.

Ruth and I had traveled the Eastern Section of the Erie Canal a few years ago on the Emita II, so my camera was idle during this part of the cruise, except to get a few more photos of the Little Falls Lock, with its lower Guillotine Gate (one of several like it in North America) and its tremendous lift — 40.5 feet. We had never traversed (or even seen) the Oswego Canal before, so I was busy with camera and data sheet at every lock. The highest lift-lock on this system is Lock #3 at Fulton, N.Y., with a "drop" (we were headed downstream) of 27 feet.

I had seen the St. Lawrence locks during construction, prior to their opening in 1959, but had never had the experience of photographing them from the "inside". They are tremendous! It was interesting also (though tedious for us, waiting) to watch a large, ocean-going freighter slowly emerging from the Cote St. Catherine Lock, above Montreal, with only inches to spare on each side. It finally cleared the lock, allowing us to proceed to Montreal. A table of data on the St. Lawrence Seaway locks is reproduced on page 49, as published in the English-language guide to the Seaway tossed on board for us at the Massena (New York) headquarters for the Saint Lawrence Seaway Development Corporation. (Eisenhower Lock).

Photo from the bow of the New Shoreham II, as we approached the unusual Guillotine Lift Gate on the lower side of Erie Canal Lock #17, (40.5-foot lift).

Some of the bridges which I photographed, from the water, included: the Hell-Gate Bridge, the Queensboro (Archie Bunker) Bridge and the 100-year-old Brooklyn Bridge at Manhattan; the George Washington, Tappan Zee, Bear Mountain, Newburgh-Beacon and Catskill Bridges on the Hudson; The I-81 International Bridge and the Three-Rivers Bridge on the St. Lawrence; and the world-famous Cantilever Bridge of 1918 above the city of Quebec.

Bill Shank

Interior of the Eisenhower Lock from the stern of the New Shoreham II, looking toward the up-stream gates. Lock dimensions: 80 feet wide by 850 long, with 42' lift.

NEW ERIE CANAL – PLUS ENERGY ISLE

Robert E. Felsburg, P.E., of Harrisburg, PA sent us this interesting article, recently published in MARINE ENGINEERING/LOG. The concept differs somewhat from the plan proposed by the Army Corps of Engineers (see AMERICAN CANALS Number 33, May 1980) but they endorse the general idea of ICONN. The "fill" from the enlarged Erie Canal would form an Energy Island Complex off the New York City harbor.

"Every other nation in the world would consider the Erie Canal one of their major assets," British-born marine engineer Nigel Chattey said recently while drumming up support for his ICONN-Erie project.

ICONN – Island Complex Offshore New York and New Jersey – and Erie projects envision creating a deepwater port and industrial and energy complex offshore, commercially and physically connected to the Great Lakes and the U. S. hinterland by a waterway system.

The complementary Erie Canal project would widen and deepen the Erie Canal so that large barges or four-barge tows could navigate the Great Lakes and cross New York State from Buffalo to Albany and down the Hudson River to the Port of New York and ICONN.

ICONN envisions construction of a new "island" complex on the sea floor on Cholera Bank alongside the Hudson Trench 15-20 miles off the coasts of New Jersey and New York, to serve as a site for energy-related activities, such as LNG, LPG, crude oil and product transshipment facilities, and eventually even for refineries and petrochemical complexes and power plants.

The island complex will provide 8,000 to 10,000 acres of land in its initial stages and include totally enclosed deepwater and barge harbors capable of handling tankers up to 500,000 deadweight tons in capacity and large Great Lakes type vessels of up to 30,000 dwt.

ICONN will cover, after completion of all its initial phases, a 4½ x 3½-mile area. Protected land for industrial plants would be set up in a series of connected polders protected by a vast series of sea dikes.

The ICONN complex will need, at maximum, 600,000,000 cu yd of rock and fill for construction.

To build this complex in such deep water presents challenges but no new technology is needed, according to Chattey who is a student of Dutch marine engineering. The vast amount of fill, he said could be provided by material dug up from the Erie Canal.

Modernization of the Erie Canal by widening and deepening it to accommodate large barges or four-barge tows would open up new commodity movements between the Great Lakes and the U. S. hinterland and ICONN and the Port of New York. Typical cargoes would be coal, grain, petroleum products, petrochemical feedstocks, fertilizers and metallic ores.

ICONN-Erie would work today, says Chattey. But, he adds, it requires a very careful and complete study before it is launched. Chattey said that Washington advised him – before applying for federal funds – to allow for a full and complete study – to get significant funds towards planning such studies from at least the states of N. Y. and N. J. (and preferably also from Connecticut and Pennsylvania) and industrial and utility users who might be potential private sector users of the island complex and/or modernized Erie Canal.

Detailed studies of the island and canal projects, emphasizing technical, environmental, legal, socio-economic and financial aspects of the idea, will take two years and cost $10 million, says Chattey.

If the study is successful, the next phase would involve detailed engineering for a portion of the project. The first dikes for the off-shore island could be built using sand and gravel dredged from the Cholera Bank off New Jersey.

"You could," Chattey said, "create the first phase of an energy island without the canal project."

To build ICONN, fill would be dropped from barges and anchored by a layer of sand, gravel and 6-ton rocks. Next would be layers of 10-ton, 45-ton and 75-ton rocks.

Two dikes would provide double protection – sheltering ships loading and unloading cargo.

A Cholera Bank port would permit access by the largest ships in the world. New York Harbor is 12.1 m (40 ft) deep. Big ships draw 27.4 m (90 ft.).

ICONN, once started, could then be developed in stages over the next 20 years, creating sites for new industry and energy-related facilities as well as for garbage and sewage disposal facilities and a deepwater port. This deepwater port would allow the importation of energy materials into the region at the lowest possible cost – something which cannot be done now because of the shallowness of the harbors along the northeast coast of the U.S.

Oil, LNG and low sulphur coal can be brought to ICONN by the efficient, available transportation systems. Low sulphur Wyoming coal or Midwestern coal can be transported by rail to the Great Lakes and by water through the Canal and down the Hudson to ICONN. Alaska crude oil can be carried by U.S.-flag tankers from Valdez around the Horn to ICONN. Grain, now barged down the Mississippi and then exported via Gulf Coast ports, could once again reach foreign lands from the Port of New York.

The U. S. Army Corps of Engineers is also studying this plan, the "All-American Transportation System," to link the Great Lakes to the Hudson via the Erie Canal. While the Corps endorses expansion of the Canal to accommodate four-barge tows, it can't, Chattey says, fully endorse the project because it overlooks the cost benefit side.

Chattey is referring to benefits to the U.S. economy of increased coal exports. This isn't gravy, either. It's a necessary step for the U.S. government to take. He points to an M.I.T. Coal Study which warns that the U.S. must increase exports of steam-coal to Europe and Japan 6-fold in the next 10-15 years to ward off Russia's planned invasions of overseas gas markets.

A dynamic speaker, Chattey is right now in the process of setting up independent ICONN-Erie councils throughout New York.

Deepening the 212.5-km (340-mile) New York State Barge Canal connecting Buffalo with Albany would revitalize Buffalo as a port and benefit Rochester, Syracuse and other cities along the canal route. This is one reason why Chattey is drumming up local support.

Barge transport is five times as energy-efficient as train transport, so, opening the canal link should lower transport costs.

"I'm not suggesting this be a toll-free canal," Chattey said. "Egypt makes a good deal of money running the State Canal. I don't see why New York State can't do the same thing."

ON THE BLACK RIVER CANAL

ACS Vice President *Bill McKelvey* sent us the above photo, made sometime around the year 1900. It shows the Delta Dam and Black River Canal Aqueduct, with a flight of three locks in the background, by-passing the dam. The Black River Canal ran 76 miles from Rome to Carthage in upper New York State, to make a connection between the Erie Canal and Lake Ontario.

TRIALS AND TRIBULATIONS

By Bill McKelvey

From Cases in the Supreme Court of New York, April 29, 1842.) ADSIT and others *vs.* BRADY.

Action on the case against the defendant for neglecting his duty as superintendent of repairs on the Erie canal, in consequence of which the plaintiffs sustained an injury. The first count alleged that the defendant, on the 29th of April, 1842, was, and ever since has been, a superintendent of repairs of the Erie canal, having section number one of the canal committed to his charge; and as such superintendent of repairs it was his duty to keep said section of the canal in good repair, and to remove therefrom all obstructions to the convenient and safe navigation thereof, so that boats and vessels usually navigating the canal might safely and conveniently, and without damage, pass upon and navigate the section. That the plaintiffs were the owners of the canal boat named Angelica, used in navigating the canal for the transportation of goods &c. That a certain other canal boat was, on the said 29th of April, sunk in the waters of the canal in the section thereof under the defendant's charge, to wit, at Watervliet, in such a position and in such a manner as to obstruct the navigation of the canal, and render the navigation unsafe and dangerous, of all which the defendant had notice; and thereupon it became and was the duty of the defendant, as such superintendent, to remove the sunken boat. Yet the defendant, not regarding his duty &c., did not remove the sunken boat, but negligently and carelessly, and contrary to his duty in that behalf, suffered the boat to remain and continue sunk in the canal from the said 29th of April until the 15th of May in the year aforesaid, obstructing and rendering dangerous the navigation &c. By reasn whereof the plaintiffs' boat Angelica, on the 6th of May in the year aforesaid, the boat being then laden with goods, wares &c., of great value, in navigating and passing along the canal at the place aforesaid, ran against and upon the sunken boat and was thereby greatly damaged, as were also the goods with which the plaintiffs' boat was laden, and the plaintiffs' boat and goods were sunk in the waters of the canal.

The defendant demurred to the declaration, assigning the following causes: 1. It is not alleged that the defendant was directed by the canal commissioners, or either of them, to make repairs on his section, or to remove the sunken boat; 2. It is not alleged that the defendant had any moneys in his hands for the purpose of making repairs or removing obstructions; 3. It is not alleged that the plaintiffs' loss was occasioned by the malicious or wilful neglect of the defendant; 4. The defendant, being an agent of the state, is not personally liable to the plaintiffs; and 5. The defendant being a sub-officer and bound to act under the direction of the canal commissioners, or one of them, he is not liable to the plaintiffs in this action.

Final Judgement

It is said that the defendant had a discretion as to what repairs were needed, and consequently that his neglect should have been charged to be wilful and malicious. But clearly the defendant had no discretion to leave this dangerous obstruction in the canal. On the facts stated in the declaration, it was his duty to remove the nuisance without any unnecessary delay.

Judgement for the plaintiffs

CANVASS WHITE — HERO OF THE ERIE CANAL (1790-1834)

Canvass White (ASCE Biographical Dictionary)

(Editor's Note: We have recently corresponded with Robert S. White II of Hackettstown, New Jersey, great, great grandson of Canvass White, Erie Canal Engineer. The following account of Canvass White activities in discovering and developing the proper "under-water" cement for the Erie Canal is provided in part by Robert White, as published in "Concrete International" *for June 1979, and also from* "Towpaths to Tugboats", *May 1982.)*

Born in Whitestown, New York, Canvass's early education included a year at Fairfield Academy, where he studied mathematics, surveying and engineering. Always in poor health, a sea voyage was recommended for him and he shipped as "super-cargo" on a merchant vessel bound for Russia. He returned in 1812, with greatly improved health, to find his country involved in a second War with Great Britain. In 1814 he organized a company of volunteers, in which he was commissioned a Lieutenant, and participated in the capture of Fort Erie. During the battle he was severely wounded.

In the spring of 1816 he was engaged by Benjamin Wright to assist in final survey work along the route of the Erie Canal. Wright and later De Witt Clinton were so impressed with the enthusiasm and natural ability of the young engineer that they suggested he make a trip to England to collect information there which would help with some of the troublesome details of the Erie. This he did, at his own expense.

He wrote his father from Liverpool in January of 1818: "I have traveled 400 miles (he later covered a full 2000), passed through a number of tunnels and over several aqueducts . . . One aqueduct which I have examined (the Pontcysyllte) conducts a canal across the River Dee in North Wales, and consists of 19 arches of cast iron . . . I shall now make a tour through the North of England."

He returned to the States in the Spring of 1818, with the latest surveying instruments, detailed drawings of English canal operations, and much information on the use of hydraulic cement in construction of English canal locks.

It was Canvass White who solved the problem of an inexpensive supply of hydraulic cement for the Erie Canal. It was not without considerable searching and researching on the part of the young engineer for a good American natural cement that would harden under water — that led him to a deposit of stone near Chittenango, New York. When burned and pulverized this stone produced the desired results. White obtained a patent for his new cement (valid in 1820) and a cement processing plant was set up, under the supervision of his brother, Hugh White, to produce this much needed material for the locks, aqueducts and bridge foundations for the Erie Canal, and later — many others.

On 18 February 1820, the Erie Canal Commissioners reported:

"The water-proof lime, which has been used, during the past season, for most of the mason work done on the canal . . . Mixed with clean silicious sand and water, and well beaten, it constitutes a mortar, which will soon set, and thoroughly cement any work of stone or brick, in which it is used under water . . . We failed repeatedly in burning, pulverising and mixing it; but many trials have now shown us the way to succeed in all these operations. And all the masons in our employ, though for some time they were loth to use it, from an opinion which they uniformly entertained of its being of no value, now regard it as a discovery of the greatest importance. It sets much quicker, and becomes stronger in the air, than common lime mortar; and under water, where a common mortar will not set at all, it begins to set immediately, and in a few weeks acquires great hardness and tenacity. It may be quarried with the same labor as common lime stone, and is known to occur in the greatest abundance, in Madison, Onondaga, Ontario and Genessee counties; its color is a yellowish grey, before it is burnt, and burning inclines it to a buff. It is softer than common limestone, and when burnt, about 10 percent lighter. It will not slack, but must be pulverized by pounding or grinding, and when reduced to powder, its bulk is not materially increased. The quantity of sand mixed with it should be about half that of the lime in bulk. From its not swelling, by being pulverized — from the expense of grinding it, and from the greatly diminished quantity of sand it will bear, it will be at once perceived, that its use will always be attended with greater expense than that of common lime. Still it may be used, at a very small proportion of the cost of any other material now known to answer the same purpose."

As work on the Erie progressed, Canvass White became the chief expert in designing its locks and other appurtenances, which earned him the title of "Principal Engineer."

Before the Erie was complete, White was approached by a number of other canal companies to assist them in their planning and design work. He became a consultant on the Schuylkill Navigation, the Farmington Canal, the Chesapeake and Delaware Canal; and was, for a time, Chief Engineer on the Union Canal, the Delaware and Raritan Canal, and the Lehigh Canal, respectively. Ill health often interfered with his canal projects, and he finally retired to Florida, where he died at the untimely age of forty-four — a great loss to the canal builders of the early Nineteenth Century.

WINTER SHUT-DOWN ON THE ERIE CANAL

By Bruce Russell

Every December New York State's Barge Canal system ceases active operations for approximately 4½ months because of freezing temperatures and the formation of ice which renders navigation either impossible or extremely impractical. Nevertheless this closure does not imply that the 570 mile waterway goes into total hibernation, with its work force laid off and a mantle of snow permitted to remain on its major structures such as locks, dams, sluice gates, retaining walls, and bridges - both movable and fixed.

Appearance of the Lockport flight from below, December 1985. The canal and locks have been drained of water, for winter maintenance. (Photo by Bruce Russell)

Instead, once the water is drained, the canal employees are shifted into other details. Winter is the time to perform needed maintenance on the 70 year old waterway, and a list of priorities is drawn up dividing the tasks into urgent and optional. The Erie Canal, or as it is sometimes called — the New York State Barge Canal - Erie Division — is no longer considered a modern waterway, and its locks and other man-made structures are beginning to show the effects of three quarters of a century of continuous use in a climate known for its extreme fluctuations, from minus 20 degrees F in winter to 90 plus in summer.

The Barge Canal's locks are constructed out of reinforced concrete rather than dressed stone blocks as in earlier canals, and during the past 10 years various sections have started to crack, crumble, and disintegrate. Water seeps into these fissures, and once the temperature dips below freezing, ice is formed which must expand and therfore additional damage is done. Larger and larger pieces then start breaking off, giving more opportunity for more water to enter and damage the section in question. Likewise the massive lock gates and the machinery to power them is not up to date or comparable to what exists on the St. Lawrence Seaway (1959) or the just inaugurated Tennessee-Tombigbee Waterway. Instead it's of the same vintage as the Panama Canal, on which construction began in 1904 during the administration of "Teddy" Roosevelt. Designed to be a modern successor to the original Erie Canal running from Albany to Buffalo, the primary motivation in constructing it was to break the monopoly of the Vanderbilt's NEW YORK CENTRAL RAILROAD within New York State along the central corridor. The thinking was that if there was no alternative means of transport that could move grain, coal, and other bulk commodities across the "Empire State" then the railroads and the "Robber Barons" who ran them could charge whatever prices they liked. Obviously at this time nobody was thinking interstate highways. Between 1904 and 1918 a new waterway was constructed roughly parallel to the old Erie Canal (and west of Rochester in the exact same alignment), and thus the New York State Barge Canal came into existence. While the earlier canal was almost totally a man-made "cut" and was hand hewn by mostly Irish laborers or "Paddys", the newer canal made extensive use of existing rivers, creeks, and streams which were widened, deepened, and regulated with dams and locks. A significant portion of the work was done with steam driven machinery, and Italian immigrants rather than Irishmen did the physical labor. Concrete instead of stone was the major building material. Shorter branches were also built to serve the Champlain and Cayuga/Seneca regions, plus a segment from Syracuse to Lake Ontario called the Oswego Canal Division.

Interior of the upper lock chamber at Lockport, drained for winter maintenance. Note the drains at the bottom of the chamber. (Bruce Russell photo)

The Barge Canal presently carries very light commercial traffic. Pleasure craft are its prime users plus an occasional inland cruise vessel such as the NEW SHOREHAM and of course Mid-Lakes Navigation Company's EMITA II. Nevertheless, everything must be kept in full working order, and winter shutdown is the only time this can occur, with the ideal months being December and March when temperatures are less brutal than in January and February. Following the final day's activity the water level in the canal is permitted to drop by means of closing certain gates and permitting no more water to flow into it from Lake Erie and other feeders. After several days all that remains in the channel is a small stream in its midpoint. Buoys and other floating navigational aids either are sitting on the dry bottom or have been previously removed and taken elsewhere. (The canal has special "buoy boats" for this purpose). The locks and their deep chambers are totally devoid of water, and this permits inspection of them and necessary repairs to be accomplished. Rust can be chipped away from areas where it has accumulated such as sluice and miter gates, red lead applied, and then final paint. Really badly deteriorated parts are noted, and a decision made on whether to replace or rehabilitate.

The most impressive locks on the New York State Barge Canal System are situated in the community of Lockport where the canal climbs the Niagara escarpment and begins the final 23 miles to Tonawanda on Lake Erie — its western terminus near Buffalo. In the days of the old Erie a series of 5 double locks was required to overcome this considerable rise in elevation, and the "Lockport Flight" was the subject of many paintings and illustrations. Traffic was so heavy at this time that the Lockport Flight actually consisted of 5 locks going up and 5 locks going down — a truly impressive sight in

(Concluded on Page 15)

WINTER SHUT-DOWN ON THE ERIE CANAL

(Concluded from Page 14)

the early part of the 19th Century. The 1904-1918 reconstruction of the Erie Canal into what was then the modern Barge Canal necessitated the removal of one of these 5 lock staircases to make room for two massive Barge Canal locks. However the other 1830 vintage flight, constructed not of concrete but of cut and dressed stone blocks, remained in use a few years strictly for smaller craft so they wouldn't have to negotiate the bigger chambers. But in the mid 1920's their lock gates were removed and the structure was converted into a spillway. Thus you have the old and the very old side by side at Lockport — truly an outstanding example of industrial archeology.

During my day at Lockport last January major rehabilitation was being done on one of the two giant locks. New concrete had been poured, and a section that had been in danger of possible collapse was being made sound again. Also one of the enormous lock gates was partially disassembled and renewal of parts was occuring. About 25 men from the canal maintenance base at Lyons, N.Y. were on the job, plus others from a private contractor. A large cement mixer had been lowered by crane onto the floor of the lock, and drills, hoses, and other tools were scattered about the completely dry floor of the chamber. Although some upkeep of the waterway is accomplished during the regular season using special work boats — primarily dredges — the principal job of keeping things in order occurs during winter shutdown, and all locks and structures are accessible by road vehicles. From time to time entire miter gates are replaced or totally rebuilt, and I am told that new ones are built either at the Lyons facility or at the canal's other maintenance base located in Waterford, near Albany. It is becoming more of a challenge to obtain working parts for the 70 year old electrical machinery that turns the locks, but the canal work force is still able to come up with ingenious solutions. Also, I have been told that European waterway people have been called in to give their expertise, and sometimes parts come from places such as Belgium, England, etc.

The canal staff is fully aware that if the waterway closes permanently they will be out of a job in an area where employment possibilities aren't good. Nevertheless among Barge Canal workers there is a great "esprit de corps" and the basic attitude is one of "CAN DO." Many of the workers I spoke to at Lockport were third and fourth generation canalers, and can trace their family roots to an immigrant who helped construct the Barge Canal in 1904. (A few may also trace their ancestry to the original Irish who dug the first Erie Canal or "Clinton Ditch" back in 1825.) New York is unique because it's the only state in the nation with a separate Department of Canals, in addition to the usual Dept. of Highways and Transportation. Morale is kept high

One of the old flight of locks has been preserved as a spillway at Lockport. Mitre gates were removed in the 1920's. The control house for the newer gates is at the lower right. (Bruce Russell photo)

by awarding a yearly prize to the best maintained, most efficiently kept lock, and many of the tenders sport T shirts with such awe-inspiring logos as "Lock 25, Best on the Erie 1983" or something similar. Some of the tenders work at this craft because it's relaxing and they're their own boss so to speak. Most view their occupation as unique in this era of obsession with automobiles to the detriment of railroads and inland waterways.

CANAL SEASON GOOD

New York's Canals, completed the second busiest navigation season of this century, as of final closing November 28th, 1986, according to Franklin E. White, New York state transportation commissioner. The Erie and Oswego canals closed November 21, while the Champlain remained open until November 28th. More than 123,000 pleasure boats locked through the state's canals in 1986, commissioner White said. There were 120,000 in 1967. Lockings in 1985 totaled 110,000. Increased use was attributed in part to Fourth of July traffic and broader awareness fostered by the 160th anniversary canal celebration last year. Commercial tug operations are expected to remain approximately at last year's level of 400,000 tons.

New York's canals remain an important commercial barge transport artery for gasoline, kerosene, jet fuel, asphalt, cement, machinery and molasses. Harnessing the canals hydropower provides a renewable source of clean indigenous energy. *(Waterways Journal)*

Future Cloudy

The long term future of N. Y. State's Barge Canal is cloudy. It presently costs $15 million a year to keep it going, which in the terms of today's bloated state budgets is miniscule. Regardless of the heroic job I saw being done on that blustery December day at Lockport, it is literally beginning to fall apart, and big, big dollars will be needed in the next few years to modernize it, as well as correct years of deferred maintenance. Stop gap measures cannot go on forever. The days of seeing a continuous procession of barges laden with grain passing through the various locks from Tonawanda to Waterford on the Hudson are gone forever, thanks to the St. Lawrence Seaway and parallel interstate highways. What might save it is the steadily increasing number of pleasure craft, plus the boom in inland waterway cruising now occuring. (See American Canals, No. 49, May 1984 article by Franz J. Katz for further details) Mini-cruise liners now proceed up the Hudson to Waterford, enter the Barge Canal, travel to Syracuse, and head into Lake Ontario using the Oswego branch of the system. Such an interesting routing would not be possible if the Barge Canal wasn't there. Other companies are planning or are building these small liners according to the aforementioned article, and no doubt will want to include the "canal cruise" in their itineries. Also for the past 13 years Mid-Lakes Navigation Company has been running moderately priced trips over the Barge Canal, and their volume has been slowly increasing. A great untapped potential exists for using this old but very beautiful and historic waterway as a vehicle to promote N. Y. State tourism. If the abandonment proposed by some factions goes through, the Empire State will have lost forever an irreplaceable asset which could do for tourism what the Rideau and Trent-Severn waterways are doing in Ontario, Canada.

In my own state of New Jersey political decisions in the early 1930s allowed the closing of the Trenton portion of the DELAWARE and RARITAN CANAL, making through journeys from Raritan Bay to the Delaware River impossible. In today's era of widespread boat ownership — when one family in eight has some type of vessel and marinas dot the landscape — such a waterway, perhaps modernized but with its historic flavor preserved, that offered a means of getting from New York to Philadelphia without venturing into the open Atlantic Ocean would be a priceless asset for the Garden State. Yet the D & R Canal was allowed to disappear as a through route, its surviving locks permanently sealed — mainly through lack of foresight, and a desire to use the right of way through Trenton for a new Route 1 alignment. Let us hope the future is kinder to N.Y.'s Erie Barge Canal so that it will forever be possible to go from the Hudson River to the Great Lakes by water.

NIGHT LIFE ON A CANAL PACKET BOAT

Canal packet sleeping berths were hung from the cabin ceiling in tiers.

We are indebted to ACS member **Monro Will** *of Rochester, New York for these gems of night travel on the historic canals of New York and Pennsylvania, as preserved by the Rochester Historical Society. The illustrations are by the late Phillip Hoffmann.*

When he was eighty-seven years of age, the late William H. Wallace, formerly a resident of Rochester, wrote his recollections of his first trip in the Erie Canal in 1846, when he was a small boy, as follows:

"My father left us at Albany after we had come up in the night boat from New York. In the way of traveling everything interested me; used to ships and steamboats, the canal boat was exciting.

"I preferred the deck, when I was allowed to sit there, watching the passing of the beautiful farm mansions and the pleasing cobblestone barns; or the gliding up to the landing place of some small village where passengers got on, or off, the boat, and where people stood offering strawberries and cherries, or cake.

"Almost every one was sent up on deck when the dinner table was being set. We were called to dinner, I suppose, about one o'clock. The table was set down the whole length of the cabin or 'saloon;' between the table and berths there was no room for waiters. It was just like a family dinner. The boat captain sat at the head and did the carving and distribution of the meats; the vegetables were arranged so that the diners could help themselves.

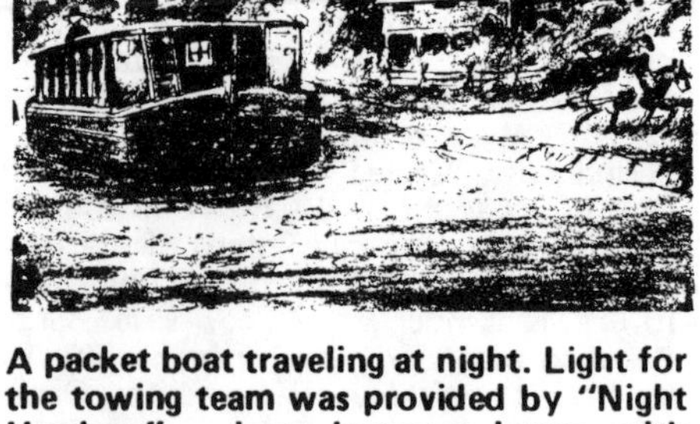

A packet boat traveling at night. Light for the towing team was provided by "Night Hawkers" — large kerosene lamps, with reflective mirrors on each side.

"After dinner, I went again on deck and was much surprised at seeing some passengers leave the boat and walk along the towpath, being taken into the boat again, about an hour later.

"Another excitement to me, were the locks. When we entered the first one, I asked what it was; a gentleman standing by told me it was the end of the canal. I soon learned about them however, and was delighted with the roaring of the water, and the boats sinking down between the stone walls, or coming up to the surface again. I was not worried by the cry 'low bridge,' and watched the people duck and double up; most of the bridges were far above my head.

"Supper or tea was something like dinner. About nine o'clock the men were either divided off by a curtain, or sent up on deck, while the women and children went to bed. I recall being highly indignant because I was put to bed in a top berth with a little girl of about my own age. However I did not go to sleep for a long time, for a company of negro minstrels stationed themselves on deck, it seemed right over my head, and danced and sang for an hour. I think these musical entertainments were not allowed to last long. I cannot recall what time it was when we arrived at Rochester the next day, but we were only one night on the packet.

"The most pleasant memory of the canal, is of our return voyage to Albany. The trip was made in a freight boat which was owned by its captain, a friend and patient of my uncle, Dr. Parson G. Shipman, who arranged the passage for us. We were the only passengers, and had the whole boat to ourselves. There was a sort of cabin in which were two staterooms. There was also a small saloon where we ate. The fittings of the boat were plain and simple. There were no damask curtains or gilded cornices, but both my mother and aunt spoke of the boat and cooking as being extremely clean and nice. The captain's wife was aboard and did the cooking, making the beds, etc. This boat was their home, at least in the summer and frieghting season. Perhaps they went down the Hudson in it to New York. We all greatly enjoyed the return trip, as we had freedom and comfort on the boat quite different from the crowded packet. It was, however, much slower and our table was not so varied; also, there were no negro minstrels. Our voyage ended at Albany, where we had just time to catch the night boat to New York."

Richard C. Ellsworth in his account of a trip "From Canton to Rochester," quotes Alvin Harlow in his "Old Towpaths," for a detailed description of canal travel. "The packet boats resembled a small Noah's Ark, a house-boat whose only deck was the roof. They were all of similar design. In the bow, cut off from the rest of the boat, was a tiny cuddy for the crew. Next back of this came the ladies' dressing room and cabin, sometimes separate, sometimes cut off from the main cabin only by a red curtain.

Drawing lots for sleeping berths on a canal packet boat.

Next was the main cabin, thirty-six to forty-five feet long, which was saloon and dining room by day, and men's dormitory by night. Back of this was the bar, and finally, at the very stern, was the kitchen, almost always presided over by a negro cook, who was usually the bartender also. The other members of the crew were the captain , two drivers and two steersmen, one each for the night and day tricks. At nine o'clock in the evening two or three members of the crew began carrying adjustable berths, sheets, pillows, curtains and so forth, into the main cabin and the passengers must go out on the deck, or huddle in the center of the room, while the berths were being made up. Each berth was a narrow wooden or metal frame with a strip of canvas fastened over it. It was held in position at one side, by two projecting iron rods which fitted into two holes in the wall of the cabin, and on the other, or front side, by two ropes attached to the edge of the frame and suspended from hooks in the ceiling. There were at least three beds in a tier, one above another, sometimes four; and all were fastened to the same rope. The mattress was a thin, flat, lumpy pad of straw, wrapped in blue canvas, and this was used on the best packets. The tiers were set as closely together as possible all around the cabin which thus furnished alleged beds for from thirty-six to forty-two people. But the boats were frequently overcrowded, seventy-five or even one hundred or more people being jammed into a single cabin."

Charles Dickens visited America, in 1842, and in his *American Notes,* published later, he made many caustic comments on the manners and customs in America. Describing his first trip on a Pennsylvania canal packet boat, he said:

"I have mentioned my having been in some uncertainty and doubt, at first, relative to the sleeping arrangements on board this boat. I remained in the same vague state of mind until ten o'clock or thereabouts, when going below, I found suspended on either side of the cabin, three long tiers of hanging bookshelves, designed apparently for volumes of the small octavo size. Looking with greater attention at these contrivances (wondering to find such literary preparations in such a place), I descried on each shelf a sort of microscopic sheet and blanket; then I began dimly to comprehend that the passengers were the library.

"I was assisted to this conclusion by seeing some of them gather around the master of the boat, at one of the tables, drawing lots with all the anxieties and passions of gamesters depicted on their countenances; while others with small pieces of cardboard in their hands were groping among the shelves in search of numbers corresponding with those they had drawn. As soon as any gentleman found his number, he took possession of it by immediately undressing himself and crawling into bed. The rapidity with which an agitated gambler subsided into a snoring slumberer, was one of the most singular effects I have ever witnessed."

Dinner on board a canal packet boat.

BIG PLANS FOR THE ERIE CANAL

The following article by Robert J. McCarthy of the Buffalo (NY) News was sent on to us recently by an ACS Member in West Palm Beach, Florida. It appeared under the heading "Big Plans for (Erie) Canal Give Crews New Pride in Their Winter Tasks":

LOCKPORT, NY - Things seem quiet during the winter around the old State Barge Canal locks.

The canal bed is drained, the locks are closed, the blue and gold state tugs are high and dry.

But in sheds at Lockport and all along the 524-mile canal system, a tradition as old as the canal itself is under way — winter maintenance.

This time of year, the veteran canal workers overhaul the motors, replace the worn cables and generally recondition the 162-year-old waterway for its rite-of-spring reopening.

"This thing has to run next year, so we have to keep at it," said Joel Byer, lock chief at Lockport. "There's no other time to do it. When we're operating, we're doing so many other things we can't stop."

Buoyed by a renewed sense of the state's commitment to the waterway that opened the American West, crews such as Byer's spend these days rejuvenating the old equipment that seems to have no place in a modern transportation world.

At Lockport, they are removing and taking apart the old electric motors that have swung open lock gates since World War I. The armatures are disassembled and baked in ovens to dry up moisture; then reinsulated, resealed and reinstalled. Control boards are polished and shined to fight corrosion, until they look as if fresh off the assembly line.

As a result of the winter ritual, Byer says, the old motors just keep on working.

Further down "Clinton's Ditch," state equipment is dredging the canal bed, a yearly chore necessitated, Section Superintendent Thomas J. Balkin said, by the silt deposited along the canal curve.

But this year, amid the old-time equipment and a century's worth of tradition, those working on the canal, Balkin says, see a new sense of optimism, a sort of renewal of purpose accompanying the state's refound dedication to the canal.

"There's been a big change in attitude around here," Balkin says. "It's finally been accepted as fact that we'll never get the big volume of commercial traffic anymore. So now, we're promoting pleasure boating instead."

Although officials of the state Department of Transportation reported an all-time low of 373,950 tons of commercial freight last year, recreational "lock-throughs" totaled about 121,000 — the second highest in the history of the canal.

"That's one of the reasons why the emphasis now is on recreation," said Barry E. Goldsberry, canal traffic agent for the department's Waterways Division.

The crumbling canal, a rich historical treasure that also serves as an important source of irrigation, drinking water and hydroelectric generation, was almost shut down by the state west of Syracuse in the dark days of 1983.

Then the department discovered itself saddled with a deteriorating system dependent on a dwindling commercial base, in need of millions of dollars worth of repairs, and no money to pay for them.

"We're talking about a major resource of the State of New York that had been neglected," Goldsberry said.

But new funds from the Rebuild New York bond issue and a federal government promise to assume 50 percent of its operation and maintenance budget finally has provided the financing needed to maintain the system.

Now, Balkin looks forward to replacing the Lockport flight of locks in 1989, marking the first such overhaul since the old Erie Canal was enlarged and modernized in 1918.

The change stems from the state's realization that the canal can contribute more to the state's economy today as a recreational and tourist resource than as a commercial waterway, leading to parks and tourist attractions on the towpath all across the state.

BY TUG-BOAT THRU THE ERIE CANAL (Part I)

The Tug and scow are shown here in a lock on the Oswego Canal. Note that the scow clears the walls of the lock by only a few inches on each side, and the end clearance between lock gates is measured in inches also. Careful manoeuvering is required both in and out of the lock.

By Gerard Chapman

It is three o'clock and I'm busy at my job when the telephone rings. It is Jeff Blinn in the New York offices of the Moran Towing and Transportation Company, calling to ask whether I can be in Troy N.Y., at three o'clock in the morning to board one of his tugs on the Barge Canal. It is a trip I have been seeking for three or four years, and I assure him that, yes, I can and shall be there.

•

Many times, in traveling on the New York State Thruway, had I seen the tranquil reaches of the Western, or Erie, Division of the New York State Barge Canal; and those who go north from Albany see occasional stretches of the Champlain Division. But so seldom is any traffic seen on these waterways that some believe them to be abandoned. Not so; about three million tons of cargo are carried yearly, with petroleum and its products predominating. (At its peak, the Erie Canal carried more than six million tons of grain, lumber, stone and other commodities annually, in the 1860's and 1870's.) I had read Walter D. Edmonds' several books about the old Erie, and had a collection of pamphlets, maps and circulars from the Department of Transportation of the State of New York about the canal, and had long wanted to see what life on it was like, as viewed from a working tug, and here at last was my wish come true.

•

It was rainy and foggy on the 70-mile drive but I arrived at Troy Lock and Dam — Federal Lock No. 1 — on the Hudson River above Albany, in good time to board the *Margot Moran* at 3:15 a.m. It was down-bound and soon reached "Troy Wall" to pick up Dunbar & Sullivan scow No. 120 moored there. I learned that it was one of two chartered to the Great Lakes Dredge & Dock Company and being moved through the canal from New York Harbor to Charlotte on Lake Ontario. The two scows had been towed up the Hudson River to Troy and No. 120 moored there while No. 121 had been pushed to a moorage just above Lock 16 near St. Johnsville in the Mohawk Valley. They were being leap-frogged through the canal to Oswego.

•

Expertly maneuvering his vessel, Mate Clayton Gillikin in the pilot house tied onto scow No. 120, lashing it to the *Margot Moran's* bow with steel cables taut as fiddle strings. Casting the scow loose from its mooring against he Wall, he turned about and started back up the Hudson, passing again through Troy Lock about 4:30 a.m., as dawn was breaking. A light rain was speckling the glassy-calm surface of the river as we reached Waterford. Here, where the Mohawk River enters the Hudson, the Erie Division of the Barge Canal branches to the west, while the Champlain Division continues north.

Captain Chester E. MacDonald, Master of the Tug "Margo Moran".

It was the first trip this vessel had made through the canal in almost two years and Gillikin observed wryly that the canal was getting to be just like the Thruway, with a sign at the junction reading: "Erie Canal, Syracuse and Buffalo; Champlain Canal, Whitehall and Lake Champlain."

•

Before we could enter the first lock on the Erie — Lock No. 2 — we had to wait while the tug *Evening Light* cleared it in downbound passage. We then entered the lock to begin the steep climb through five locks in close succession to ascend 169 feet around the cataracts over which the Mohawk tumbles into the Hudson. They are situated on only two miles of canal and provide a spectacular entry to the waterway to the west. Above Lock 6, at 184 feet above sea level, the canal extends west for 10 level miles, and we can pause to contemplate this great inland waterway.

Designated No. 1 on the Erie Division, the lock at Troy, at the head of tidewater, is controlled by the Corps of Engineers, U.S. Army, but all other locks on the system (except one near Buffalo) are under the jurisdiction of the Department of Transportation of the State of New York, which also is responsible for its motor roads.

•

The *Morgot Moran* is only one of many tugboats operated by the Moran Towing and Transportation Company, with headquarters in New York but operating in several ports on the East, Gulf and West coasts of the United States. Most of the time they are engaged in the prosaic business of docking or undocking large ships, and towing scows or barges in the bays, rivers and canals carrying our water-borne transportation. Sometimes they are called upon to rescue vessels in distress upon the high seas, victims of storms or mechanical failure, but tales of their dogged heroism cannot be told here.

Adapted for canal work, tugs such as ours have retractable pilot houses, and it was startling when we encountered the first low bridge, to have the pilot house suddenly drop 10 or a dozen feet so that we could pass under it.

•

During our transit of the five stair-step locks, as we climbed out of the Hudson Valley into the valley of the Mohawk, the watch changed at 6 a.m. The mate relinquished the controls to the captain, and the rest of the crew changed also, as those who had been sleeping relieved those on duty. All except the cook work six hours, then are off six hours. Assigned to share a cabin with the mate, I avoided entering the cabin during his rest periods, and it was not until early afternoon, when he was again on duty, that I lay down at last for a little rest.

In little more than an hour we reached Lock 7, and with eleven miles to go, it was the better part of two hours to Lock 8, as the tug with its burden goes between seven and eight miles per hour. Capt. Chester E. MacDonald was in the pilot house. A veteran of many years on Moran tugs, he wore an air of quiet, competent authority and the assurance of being completely in control of the operation. Of medium height and build, with white, close-cropped hair, he was affable and responded to my questions with stories of his years of experience.

The rain had stopped, the sun was shining, and there was no wind. As the miles unwind, the skipper reminisced about life on the canal in the past years, telling how the towboat companies used to vie to be the first through the canal in the Spring. "I remember one time, about 20 years ago," he recalled, "when several of us were waiting below a lock which was clogged with the accumulated debris of winter: rubbish, limbs of trees, logs. The engineers were having a hell of a time clearing it. I knew the lockmaster and offered to push the stuff out of his lock if he could manage to admit my tug to it. He agreed readily, since it would get them out of a tough spot. In two or three careful nudges, I pushed the solid jumble of debris up out of the lock and over toward the dam, where it was carried away by the current. And, of course, that put me first through the lock, ahead of all the others, and I kept ahead all the way up!"

•

Passing through a lock involves the cooperation of deck hands and pilot. Upon nearing a lock, the pilot cuts his engine and drifts to a very slow approach. If the gate is closed, he sounds three short blasts on his whistle to request passage. If the gate is open and the signal green, he enters the lock. Meanwhile, when the deckhands hear the engine go down, they drop whatever they may be doing and scramble over the bow onto the scow, to stand one on each side at its bow, to signal the pilot to go right or left to center it in the lock. For there is only a foot and half difference between the width of the scow and that of the lock — a tight fit for a float so long and heavy. When in the lock, they cast their heavy lines onto snubbing posts to prevent the float's drifting into the gates.

The gate behind the two is closed. If the lockage is from a lower level to a higher, water is admitted to the lock from the canal above, filling it to the higher level. As the float rises, the deckhands cast their lines onto successively higher snubbing posts set into the walls. If the lockage is from a higher level to a lower, the reverse procedure is followed. As the float descends, the deckhands flip their mooses from the snubbing posts, but when the line fails to come free, one of them has to climb the ladder set into the wall of the lock to free it, and climb down again. Sometimes small boys watching the operation do it for them.

At the new level the lines are cast off, the float moves slowly into the canal and conversations with those standing on the walls fade out.

At the first lock on the Erie, we obtained Clearance No. 90, and in each subsequent lockage that number was given to the locktender for his records.

Water for filling the locks comes out of the canal, which in turn is supplied either by intersecting streams or by aqueduct from the distant Catskill Mountains. To conserve water, small boats are required to wait for scheduled lockages, but when we ran light (without a scow) several of them entered the locks with us to get through sooner.

Frank Schoonmaker, The Cook.

The Tug "Margo Moran" in one of the Locks on the Erie Canal.

Between lockages, the deckhands have other duties: they pass and tighten the cables used to bind the scow to the bow of the tug; at night, on the forward end of the scow they place red and green signal lamps, and a white one in the center for the guidance of the pilot in steering; and they do maintenance work during the long runs between locks.

As the hours wore on and we ascended the canalized Mohawk River, only one problem arose: the engineers had lowered the level above Lock 8 to work on its dam, and the pool between it and Lock 9 was shallow, requiring a slower rate of travel. And since our scow drew 11 feet of water, we barely scraped over the sill at lock 9.

For the greater part of its way to Rome, the Erie Division follows the Mohawk River in its meanderings and the scenery is varied and beautiful. Its aspect is mostly pastoral, with an occasional city or village for a change; fields of growing crops alternate with pastureland dotted with grazing cattle. Islands dot the course of the river, with buoys to guide the navigator through the deepened channel. On the upbound run, white buoys bearing odd numbers are on the left, with even-numbered red buoys on the right. When night falls, photocells turn on their lights, which flash briefly at intervals, white or red. One's location at any moment can be determined by reference to Lakes Survey Chart No. 180, which depicts every buoy or stationary light by number. Frequently visible on one side of the canal or the other, is the Thruway, its speeding traffic in sharp contrast to our plodding progress.

•

At noon the Mate reappeared, refreshed, and boomed jovially, "Okay, Skip, I'll take her now. Anything new?"

"No," answered MacDonald, "just be sure to let the office know where we are late in the afternoon," and with that he descended for his rest.

He was referring to the requirement that the Moran headquarters in New York be informed at all times where their many vessels are. During his watch the Captain had used the radiotelephone to call the dispatcher there, Mike Bodlovich. In his turn, the Mate used a lockside telephone to call in.

Sometimes, on another channel, the radio blares as a pilot somewhere on the canal talks with another or to canal personnel.

Late in the day, after locking through No. 16, we passed scow No. 121 at its mooring in the basin and went on.

(To be Continued)

"CANAL BOATMAN"

"Canal Boatman" by Richard Garrity was written originally in 1977 and has been re-issued in paper-back form with a foreword by Lionel Wyld. The author refers to it as "My Life on Upstate Waterways" and tells in considerable detail about his experiences as a boy on his father's boat on the Erie Canal, and later, as a husband and father striving to raise a family, working the Barge Canal and other waterways. 240 pages, 60 photographs, as well as a map of the entire upper New York canal system, old and new. $12.95 from Syracuse University Press, 1600 Jamesville Avenue, Syracuse, New York 13244. Phone (315) 443-2597. Add $2.00 for postage and handling.

The New York State Canals completed 164 consecutive years of service on November 29, 1989, concluding a season of 213 days. The number of pleasure craft raised or lowered from one level to another during the season totaled 159,141. Commercial cargo shipped on the system totaled 345,735 tons.

PLEASURE CRAFT SUMMARY

CANAL	1989
ERIE	93,353
CHAMPLAIN	26,756
OSWEGO	24,301
CAYUGA-SENECA	14,731
TOTAL	159,141

(From "The Canawler Barge", Summer 1990)

"A magnificent boondoggle" is how industrial archaeologist Thomas Flagg described the 2-million-bushel grain elevator at the foot of Columbia St. in Brooklyn. The huge facility was built in 1922 to accommodate grain shipments arriving from the Great Lakes region via the New York State Barge Canal. It was never used to anywhere near capacity, and ceased to operate altogether in 1965, but remains a dominant and still magnificent feature of the landscape, according to an article by Christopher Gray in the New York Times of May 13, 1990. (Contributed by both Charles F. Hruska and Walter L. Meseck)

BY TUG-BOAT THRU THE ERIE CANAL (Part II)

We pass a work boat and barges of the New York Department of Transportation on the Erie Canal. Note the dredging equipment towering above the trees on the bank.

By Gerard Chapman

After we passed through Lock No. 17, with the highest lift on the system, I finally went to bed while the mate drove on through the night from midnight to six in the morning.

My bunk, about three feet wide, was the upper of two in the Mate's cabin, and only a steel bulkhead separated it from the engine room. The pounding of the engine was heard and felt throughout the vessel, and especially severely in that confined space, but after more than 40 hours of wakefulness, I was tired enough to sleep fitfully — and by the next two nights I had become sufficiently acclimated to its presence to sleep well.

FRIDAY

When the cook rang his bell at 5:30 and I prepared to vacate the cabin for the Mate when the watch changed at six, we were in the summit level (420 feet above sea level) beyond Lock 20. In the basin at Rome we moored the scow and, turning about, ran back down through the locks to pick up the other scow just above Lock 16.

This time we went through No. 17 by daylight. At Little Falls, it is situated in a rocky gorge and from a distance suggests a fortress high on a hill. It has the highest lift in the system which, at 40½ feet, exceeds that of any lock on the Panama Canal. It also has the only vertically travelling, or lift, gate (also called a "Guillotine Gate"). Seen from within, the lock is a vast cavern with sheer concrete walls towering far above the float.

•

All the locks on the Barge Canal are 328 feet between gates, with an available length in the chamber of 300 feet; width is 45 feet, with 12 feet of water over the sills. Gates are of the mitre type, akin to a pair of casement windows. Barges and scows are about 43½ feet wide, calling for a high degree of skill on the part of the pilot to guide it into the lock without undue bumping and scraping. The combined barge and tug — the float — is about 300 feet long and care is necessary to avoid ramming the gates. Both our pilots — Capt. MacDonald and Mate Gillikin — had light touches indeed upon the controls and were experts at "threading the needle." When entering a lock they contend with hazards such as cross currents from the adjacent dam, a curved approach to a lock, and high winds. When I expressed admiration for his skill, Gillikin said, "All that I know about this business I learned from the Skipper, and we've been together on this vessel for many years."

It was evident that each had complete confidence in the other.

Back up the canal with scow No. 121 we went, passed the other one moored at Rome, and began to descend from the summit level at Lock 21. This portion of the Canal, as with the summit level, no longer followed the Mohawk River but was a dug channel with few curves, no islands, and a regular shoreline. It was, accordingly, much less interesting in scenic attributes and rather monotonous. The Mohawk River wandered in and out, entering at one side and within a short distance, leaving over a spillway on the opposite side.

The undeviating straightness of the canal, its regular shoreline and its constant width gave rise, however, to an interesting phenomenon not observed elsewhere: standing waves which pursued the float at equal speed with it. Directly abeam of the tug the water would recede from the shore, exposing the rock rip-rap momentarily, and then a surging wave of cresting white water would inundate the rocks far above the normal water line. Birds would often fly onto the exposed rocks, looking for food, then have to spring out of the way of the overwhelming wave.

In both canalized river and dug portions of the canal one frequently sees the yellow-and-blue dredges of the Department of Transportation of the State working to keep the channel at its prescribed depth and free of obstructions. Hydraulic dredges draw muck from the bottom, pumping it through large pipes supported on pontoons, to low-lying land bordering the canal; the water runs back into the canal, leaving an alluvial deposit of fertile soil which the State leases to farmers for crops. Alternately, clamshell buckets scoop rocks or muck from the channel, which are conveyed by barge for dumping elsewhere.

The summit level and beyond was home territory to Capt. MacDonald, who lives in Rome. His wife and other relatives and friends were at Lock 20, 21 and 22 to see him and his tow, since this was his first trip up this way in almost two years. The conversations about family and friends in the soft summer evening made it seem like a homecoming celebration.

•

Upstate New York is also familiar territory to Frank Schumaker, the cook, whose home is in Watervliet. An affable bachelor, he devotes a lot of his off-duty time taking neighbor boys on camping trips. On board the tug, he sets a marvelous table. For breakfast he offers hot oatmeal and several dry cereals, eggs cooked to order, bacon or sausages. Lunch is filling, with meat or fish, potatoes, two vegetables and dessert, and at dinner he offers a choice of meat or chicken or fish in ample quantity, and all the fixings. On Sunday he always serves turkey. The refrigerator in the galley is stocked with cold meat for sandwiches for between-meal snacks, and with a variety of soft drinks and milk. The crew eats well — and so did I. Frank cooks on a stove which burns diesel fuel; the fire burns low between meals to keep a pot of coffee warm.

As twilight dimmed the sky, the undeviating canal stretched into the distance like a broad silver ribbon tinged with pink. Finally the winking signal lights were the only indicator of the route ahead. Then the pilot switched on the searchlights; one mounted on the right side of the scow limned the adjacent shoreline while the other, atop pilot house, was focused on the left bank. Without diminishing speed, the float surged on through a seeming canyon of brightly lighted trees within an illimitable blackness.

The bright lights of Lock 21 and of 22 soon after, loomed out of the darkness, had their brief moments as we locked through, and faded into the rear. The watch changed and I slept during the transit of Oneida Lake and beyond.

•

SATURDAY

When Frank rang his bell at 5:30, we had gained Lock 1 on the Oswego Canal. We were still going down hill and would continue until reaching the level of Lake Ontario, at 244 feet above the sea, the lowest of the five Great Lakes.

This canal, branching from the Oneida River beyond Lock 23, follows the course of the Oswego River. There are seven dams with associated locks, numbered from 1 to 8; there is no Lock 4. The aspect of the canal seems quite different from that of the Erie; even its rural stretches seem more built up, with waterfront homes and boats. Following the river, it too twists and turns and has islands, but church spires and factory chimneys appear at the far end of a reach. Much of the canal is wedged between Niagara-Mohawk powerhouses situated at the dams, or factories on the river banks. All give the impression of great age — of being holdovers from the nineteenth century. The constricted lockside areas do not permit the spacious green, manicured lawns of the Erie; there is barely room for the lockhouses. And overhead there are many more railway and highway bridges to cause the pilot to run with his domain retracted into its shell much of the time.

Finally, in close succession, we descended through Locks 5 to 8 to emerge upon the harbor at the Port of Oswego on Lake Ontario. There we moored scow 121 alongside the Port Authority wharf and started back up the canal.

On this passage we crossed Oneida Lake in the late afternoon under bright skies and with a following wind. Under full throttle the tug outran the waves, plunging her bow into the big ones and throwing green water far to either side. An oil barge, *Morania 300,* coming the other way and breasting the waves, threw white spray high enough to obscure itself momentarily.

•

With a 21 mile run from Brewerton at the western end of the lake to Sylvan Beach on the east, the Mate set the tug on automatic pilot and sat back to take his ease. "Watch how well she holds the course," he suggested. "Even though the tail wind tends to make us yaw, the automatic pilot will compensate and keep us on course within about three degree either side." And so it was; the gyrocompass, which both indicates the bearing and actuates the corrective action of the automatic pilot, showed almost no deviation from the desired course.

When navigating the narrow canals, the human pilot seldom takes his hand from a short steering lever, as it requires constant small adjustment to keep the long tow aligned with the axis of the waterway. Susceptible of very small adjustment by the Sperry non-follow-up steering system, the hydraulically actuated, large rudder exerts virtually instant effect upon the heavy tow. Only a small deviation from the straight-and-narrow could run the float onto a shoal or into the mud. In his six hours of duty the pilot must make literally thousands of tiny corrections to keep out of trouble. And so he welcomes an opportunity such as offered by Oneida Lake to avail himself of the respite afforded by the Sperry Automatic Pilot, which works in conjunction with the steering system but which deactivates the steering levers. He can monitor the setting of the rudder from a gauge showing its angle in degrees right or left.

Although keeping his eye upon the buoys marking the course across the lake, Clayton Gillikin could relax and talk. His speech is mildly Southern and responding to my query, he said: "I live in Moorehead City, N.C. My wife and I operate a small bar and lunch counter, which she tends while I'm away. I fly down for my two weeks off and fly back again and in either place the change in work is a welcome variation. My wife and I like to travel and we have a camper. We hope to get enough time to drive up north — up here — and see some of the country beyond the canal. Our little girl is old enough to travel now and we'd like to show her something new and different."

The "Margo Moran" heads out of Oswego Harbor, with one of the two barges, which it "leap-frogged" all the way from the Hudson River. After this photo was made, the crew picked up the other barge in the outer harbor and headed west on Lake Ontario for Rochester.

Lock Number 17 on the Erie Canal at Little Falls, at 40-½' lift, is the highest lift lock on the entire New York State canal system. An interesting feature of this lock is the lower "Guillotine Gate" which rises overhead to permit boats to pass in and out of the lock underneath it.

•

A big man who habitually wears a visored cap set rakishly on his head, Gillikin pointed out the other instrumentation and controls. A gauge indicates propellor revolutions forward or backward; there is a magnetic compass; a large lever controls the raising or lowering of the pilot house; there is a radar set, unused on this trip. Buttons can start or stop the engine without intervention by the engineer. For so small a vessel, the *Margot Moran* possesses an imposing array of modern controls and instrumentation appropriate to navigation on all types of water: narrow canals, broad bays, and open lake or ocean.

Out of the Oswego and into the Erie we proceeded, ascending to the summit level again and about 7:30 p.m. we tied onto scow 120 at Rome. Then it was back down through the now-familiar route, affording a chance for Bob Volz, Chief Engineer, to show me his domain below decks.

His engine room was spotless, with "a place for everything and everything in its place." Wearing a pair of acoustical earmuffs I could understand him quite clearly despite the din of the laboring engine. Bob was proud of his Fairbanks-Morse machine developing 1,280 horsepower and one of only a few made, since it was not a popular design. With eight cylinders it has 16 opposing pistons instead of cylinder heads and most engineers consider it too complicated. But Bob finds it easy to maintain, and he got his job with Moran because he is an expert on that type of engine.

He pointed out all the auxiliary apparatus: steering gear, electrical generator, batteries for use when the engine is shut down; compressed air tank for starting the engine; pressurized tanks of drinking water and ordinary water from the lake or canal; the electrical distribution panel; fuel tanks with capacity for a week's operation or longer; and other essential components of a self-contained floating power plant and residence for nine men.

Alternating six-hour watches with Ben Larson, Assistant Engineer, Bob Volz admitted that life during a tour of duty was exacting and confining. "But," he said, "it's a good life. I like the work and Moran is a good company to work for. We are well paid and, actually, with two weeks on and two off, we work only a half a year. Where most shore-bound men work a whole year for a two-week vacation, I have two weeks off every month. I live in a waterfront house in Florida and have a boat, with time to use it. I fly home and fly back to work, wherever the vessel might be at the time. The scenery changes, and the air fare is deductable!"

Charged with general maintenance of the tug, Volz oversees all the equipment, and had been chipping rust during the trip so that the deckhands could apply fresh paint to keep the boat looking trim.

It was evening in Sylvan Beach as we approached the lake for the third time. The lights were bright and the Saturday evening crowds all but surrounded us as we floated past; we seemed almost a part of that traditional weekly convocation. We started across Oneida Lake about 11 p.m. under the light of a brilliant, almost full, moon which cast a long silvery track upon the rippling water. The glow of Sylvan Beach faded astern, lights winked all around the horizon and we throbbed along under the guidance of the automatic pilot. When the watch changed at midnight I reluctantly left the enchanting scene and retired for the night.

SUNDAY

Down through the succession of locks in the Oswego Canal we came to the harbor once again, and there I left the tug to return home. While waiting for the bus I climbed to old Fort Ontario, overlooking the bay, and saw the *Margot Moran* stand out into Lake Ontario past the breakwater, turn west while towing the two scows in tandem, and on its final lap, head toward Charlotte, port for Rochester.

We are happy to welcome into the ranks of LIFE MEMBERS of the American Canal Society: *Arthur Sweeton III* of Canton Center, Connecticut and *John P. Grossman* of North Tonawanda, New York. This now brings total membership in this special group to sixty-three.

SONGS OF THE "HORSE-OCEAN SAILOR"

By Dr. William R. Hullfish

Dr. Hullfish, Professor of Music at the State University of New York, College at Brockport, is also a writer and singer and leads his own canal-ballad singing group known as the "Golden Eagles String Band." He is the author of THE CANALLER'S SONGBOOK, recently published by the American Canal and Transportation Center.

Come all ye dry land sailors and listen to me song,
It's got a thousand verses but I won't detain ye long,
It concerns the adventures of this Sou' Lisburn barge,
As I sailed as a man before the mast on the good ship Calabar.

Before the drone of motors drowned out the singing, songs echoed up and down the length of America's canal system. These songs tell us much about the nature of the canaller's work, his attitude toward himself, and the attitude of others.

Because the nature of the canaller's work was not one of hauling in anchors, lowering boats over the side, raising sails and hauling up yardarms, canal songs reflect more of filling in the slack time rather than hauling in slack sails. Unlike sea shanties, the canal repetoire contains almost no work songs. There is nothing resembling capstan shanties, long and short haul shanties, or other sailing work songs.

Using Borrowed Melodies

There are also very few canal songs with original melodies. Unlike sea shanties, whose melodic form comes from the type of work being performed, canal songs take their melodies and form from the popular music of the day.

Canallers apparently whiled away much of the time between towns and locks by making up personal songs to familiar melodies. Songs with dozens of verses, detailing a particular canal, are more common than universal songs adopted (and adapted) by generations of sailors. Canal songs have an individual nature about them rather than the universal, group singing approach.

Even canal songs that do contain a common chorus, which might have been sung by a group, often change the words to each chorus — a practice which discourages group singing. Many of the canal songs with singable choruses may not have been sung by canallers at all, but by professional singers.

Take this nice singable, call and response chorus from "Oh! Dat Low Bridge."

Look out, dat low bridge,
(Look out, dat low bridge)
Look out, dat low bridge,
(Look out, dat low bridge)
The captain, cook and all the crew,
Oh, duck your head way down,
The fastest ship in all the fleet,
Two Sisters come to town.

Even though the call and response form is used in many work songs, in this case it is more typical of minstrel show music. "Oh! Dat Low Bridge" was a song from Braham and Harrigan's 1880 minstrel show, The Grip.

A Typical Canal Song

More typical of canal songs is the popular epic song of the Nineteenth Century, "The Raging Canal." This song has no chorus, no call and response refrain and requires the singer to remember dozens of verses. In short, a song meant to be sung by one person.

Verse 1:
Come listen to my story ye landsman one and all,
I'll sing to you the dangers of that raging canal,
For I am one of many who expects a watery grave,
For I've been at the mercy of the wind and of the wave.

Verse 4:
The Captain told the driver to hurry with all speed,
And his orders were obeyed, for he soon cracked up his lead,
With the fastest kind of driving, we allowed by 12 o'clock,
We'd be in Schenectady right bang against the dock.

Verse 13:
Now if I live a thousand years, the horrors of that night,
Will ever in my memory be, a spot most burning bright.
For nothing in this whole wide world will ever raise my gall,
Except the thoughts of my voyage on the raging canal.

Although there is no poroof that "The Raging Canal" was ever sung by canallers, it is very similar to songs collected from canallers, such as Captain Pearl R. Nye's epic, "The Old Canal." Both songs use borrowed tunes. "The Raging Canal" is based on "Caroline of Edinboro Town" and "The Old Canal" borrows the tune to "The Little Old Log Cabin in the Lane." Captain Nye's song has seventy-six verses and documents the entire geography of the Ohio-Erie Canal.

Solo Versus Group Songs

Instead of songs to coordinate a group effort, the canallers used more solo songs. Unlike large sailing vessels, the canalboat had a small crew and jobs were individual, rather than group chores. The driver walked along the towpath and the steersman stood at the tiller.

It must have been common practice to take a familiar ditty, like "Pop Goes the Weasel," and make up verses to fit the particular canal. In this instance, the canal was the Delaware and Hudson.

Round and round the Wurtsboro bend,
The big boat chased the squeezer,
Pat Flax's boat had passed them both,
Slicker than a weasel.
Slicker than a weasel.

The subjects used in canal songs tell us a bit about how the canaller viewed himself and how he was viewed by others. Songs about sailing were, quite obviously, very popular. The mule was also a popular subject and the most sung about crew member was the cook.

Sea Songs

Canallers adapted a number of famous sea songs for canal purposes. True to form, they chose solo songs. "The Dark-eyed Sailor" was easily adapted by changing the word "sailor" to "canaller." Captain Nye changed "The Jolly Skipper" in one sailing song to "The Jolly Boatman." Nye also changed the location of the skipper's exploits from the sea coast to Akron, Ohio.

At least two of Henry Russell's famous sailing songs were adapted to the canal. "A Life on the Ocean Wave," later to become the official song of Royal Navy, became "A Life on the Raging Canal." Instead of the line, "A life on the ocean wave, a home on its rolling deep," it became, "A life on the raging canal, a home on its muddy deep."

One of the reasons for canallers using sailing songs and references to sailing may have been to convince others that they were true sailors. Apparently many townspeople along the canals thought of canallers as little more than gypsies. Most canals were closed for at least a few months of the year so canalling was not even looked on as a full time occupation.

When I asked a 90-year-old resident of the canal town of Brockport, New York, if he remembered any canal songs, he proceeded to chant the following words to a sing-song melody.

Canaller, canaller, you'll never get rich,
You work on Sunday, you'll die in the ditch,
Canaller, canaller, you son-of-a-bitch.

Canallers also had to defend their profession against the "open water" sailors who referred to canallers as "horse ocean sailors." This shows up in such lines as:

I'm every inch a sailor boy, on the E-ri-e Canal."

There were any number of songs that made fun of the "dangerous" life of the canaller. In these songs, the canalboat crew hung on for dear life during violent storms, never realizing that safety was only a few feet away.

The captain came on deck with a spy glass in his hand,
But the night it was so dark he could not discover land,
He said to us with a faltering voice, while tears began to fall,
"Prepare to meet your death this night on the raging canal."

In Mark Twain's parody of "The Raging Canal," published in his book, *Roughing It,* he describes the rescue as follows:

For straight a farmer brought a plank,
mysteriously inspired,
And laying it unto the ship,
in silent awe retired.

Then every sufferer stood amazed,
that pilot man before,
A moment stood, then wondering turned,
and speechless walked ashore.

An Irish canal song, "The Good Ship Calabar," ends in much the same way.

Then we all fell into the water,
and we all let out a roar,
There was an old farmer
standing on the edge of the bank,
And he threw me the end of his braces,
And he pulled me into shore.
No more I'll be a sailor
and go sail the raging main,
And the next time I go Lisburn,
I'll take the bloody train.

Mule Songs

Before motors took over, canallers sang about their mules, or in some cases, horses. Not all of the songs were complimentary. One song, a parody on the minstrel song, "Never Take the Horseshoe from the Door," warned canallers to "Never Take the Hindshoe from a Mule."

The business end of a mule is mighty ticklish,
Never, never touch him as a rule.
He'll kick you full of holes in seven seconds,
Trust him not there's mischief in a mule.

The kicking mule must have been notorious because he found his way into many songs.

Now this mule he was a daisy,
He pulverized the hog,
dissected seven chinamen,
and kicked a yellar dog.
He kicked as quick as lightning,
And had an iron jaw,
He's just the mule to have around,
to take your mother-in-law.

Captain Nye made up a song about one of his mules that was "cured" of kicking.

He was in the army and there you obey,
But he was like others, got smart by the day,
He sure would act stubborn would balk, kick and reel,
And want to kick you at most every meal,
So all became tired at last of his pranks,
For when he felt like it, would run, break the ranks,
He'd bray like a wild ass on mountain, in vale,
'til a surgeon got wise, cut the nerve in his tail.

Some songs told about the ailments and diseases common to mules.

When I drove a team of spavined mules,
On the E-ri-e Canal.

Some canallers bragged about their team.

Attend all ye drivers, I sing of my team,
They're the fleetest and strongest that ever was seen.
There's none that will toil with such speed down the creek,
Or start at the word of the driver so quick.

Songs About the Cook

Of course, of all the crew, the cook was the most sung about and also the most insulted. The poor cook couldn't do much of anything right. Her cooking was terrible, her disposition was even worse, and her appearance and age were the object of scorn.

The cook she's a daisy, she's dead gone on me,
With her fiery red hair and she's twice twenty-three,
She's crosseyed and freckled, a dumpling and a pet,
We use her for a headlight at night on the deck.

and

The cook we had on board the deck,
stood six feet in her socks.
Her hand was like an elephant's ear,
and her breath would open the locks.
A maid of fifty summers was she,
Most of her body was on the floor,
and when at night she went to bed,
Oh, sufferin' how she'd snore.

One thing that sets canal songs apart from other repertoires is the ever present good humor. Despite bad weather, bugs, insults and long delays at locks, the canallers laughed at it all, and most of all, themselves.

Like a fish on the hook I pine,
On his dull unchanging shore,
Oh, give me the packet line,
And the muddy canal's dull roar.

or

I've traveled all around the world,
And Tonawanda, too.

or

With the beef steak as tough as a fightin' dog's neck,
And the flies playing tag with the cook on the deck.

or

There's gravel on the towpath, hornets in the sand,
Oh, pity poor canallers that's far away from land.

or

The cook she wrung her hands and she came upon the deck,
Saying, "Alas what will become of us, our boat it is a wreck,"
The steersman knocked her over, for he was a man of sense,
And the bowsman jumped ashore and lashed her to a fence.

Because of the individual nature of the canaller's work and there being no necessity to coordinate the work of a large crew, the canal song remained a personal, solo song. The motor drowned out canal singing early in the century so most of the personal repertoires of canallers were lost.

We are fortunate that a few individual collections have been handed down in families or recorded before their authors died. James Lee, author of *Tales the Boatmen Told,* published three or four songs handed down in Captain Peter Lenstrohm's family.

John Lomax collected and recorded the songs of Captain Pearl R. Nye just prior to World War II, and only months before Captain Nye died, Ohio State University recorded over thirty of his canal songs. However, this is the exception rather than the rule. Untold numbers of canal songs died along with the canallers who made them up. The typical final verse of many canal songs has a ring of truth.

Now I'm the only son-of-sea cook,
That's left to tell the tale.

Bibliography

Hullfish, William. *The Canaller's Songbook.* York, Pa.: American Canal and Transportation Center, 1987.

Lee, James. *Tales the Boatmen Told.* Exton, Pa.: Canal Press, Inc., 1977.

Nye, Pearl R. *Songs and Correspondence.* Microfilm 12, Washington, D.C.: The Library of Congress, 1937.

Thompson, Harold. *Body, Boots and Britches.* Philadelphia, Pa.: J.B. Lippincott, 1939.

Wyld, Lionel. *Low Bridge: Folklore and the Erie Canal.* Syracuse, N.Y.: Syracuse University Press, 1962.

Recordings of Canal Songs:

Body, Boots and Britches. Folkways Record FTS32317.

Grand Canal Ballads. Folkways Record FTS32318.

Library of Congress. LC Tape #72 (Lomax recordings of Captain Pearl R. Nye. Some are canal songs.).

Ohio State University, Columbus, Ohio. Cassette M-10 (about 30 canal songs by Captain Pearl R. Nye).

Oh! That Low Bridge. Front Hall Record, FHR-028.

Songs of Ohio Country. Mark Record, MC1035.

ACROSS NEW YORK BY TRAIN AND CANAL – 1837

by William Dzombak

Captain Frederick Marryat retired from the British navy in 1830, at age thirty-eight; he then took up writing, and authored several successful novels. In May, 1837 he arrived at New York and, after touring America, returned to England in November, 1838. The following passages have been taken from a diary he kept while in America.

He travelled from Schenectady to Utica by railroad, and left us the following description of the train ride.

"There is one disadvantage generally attending railroads. Travellers proceed more rapidly, but they lose all the beauty of the country. Railroads of course run through the most level portions of the States; and the levels, except they happen to be on the banks of a river, are invariably uninteresting. The road from Schenectady to Utica is one of the exceptions to this rule: there is not perhaps a more beautiful variety of scenery to be found anywhere. You run the whole way through the lovely valley of the Mohawk, on the banks of the Mohawk river. It was really delightful, but the motion was so rapid that you lamented passing by so fast.

"The Utica railroad is one of the best in America; the eighty miles are performed in four hours and a-half, stoppages for taking in water, passengers, and refreshments, included. The locomotive was of great power, and as it snorted along with a train of carriages a half mile long in tow, it threw out such showers of fire, that we were constantly in danger of conflagration. The weather was too warm to admit of the windows being closed, and the ladies, assisted by the gentlemen, were constantly employed in putting out the sparks which settled on their clothes - the first time I ever heard ladies complain of having too many sparks about them. As the evening closed in we actually were whirled along through a stream of fiery threads — a beautiful, although humble imitation of the tail of a comet."

Captain Marryat rode the Erie Canal from Utica to Syracuse and then the Oswego Canal north to Oswego, on Lake Ontario, as he tells us next.

"Set off for Oswego in a canal boat; it was called a packet boat because it did not carry merchandise, but was a very small affair, about fifty feet long by eight wide. The captain of her was, however, in his own opinion, no small affair; he puffed and swelled until he looked larger than his boat. This personage, as soon as we were under weigh, sat down in the narrow cabin, before a small table; sent for his writing desk, which was about the size of a street organ, and like himself, no small affair; ordered a bell to be rung in our ears to summon the passengers; and then, taking down the names of four or five people, received the enormous sum of ten dollars passage-money. He then locked his desk with a key large enough for a street-door, ordered his steward to remove it, and went on deck

Captain Frederick Marryat, about 1836, by S. A. Mount. (Metropolitan Museum)

to walk just three feet and return again. After all, there is nothing like being a captain.

"Although many of the boats are laid up, there is still considerable traffic on this canal. We passed Rome, a village of two thousand inhabitants, at which number it has for many years been stationary. This branch of the canal is, of course, cut through the levels, and we passed through swamps and wild forests; here and there some few acres were cleared, and a log-house was erected, looking very solitary and forlorn, surrounded by the stumps of the trees which had been felled, and which now lay corded up on the banks of the canal, ready to be disposed of. Wild and dreary as the country is, the mass of forest is gradually receding, and occasionally some solitary tree is left standing, throwing out its wide arms, and appearing as if in lamentation at its separation from its companions, with whom for centuries it had been in close fellowship.

"The mosquitos of this district have reaped some benefit from the cutting of the canal here. Before these impervious forest retreats were thus pierced, they could not have tasted human blood; for ages it must have been unknown to them, even by tradition; and if they taxed all other boats on the canal as they did ours, a canal share with them must be considerably above par, and highly profitable . . . .

"The first sixty miles of this canal, which is through a flat swampy forest, is without a lock; but after you pass Syracuse, you have to descend by locks to the Oswego river, and the same at every rapid of the river; in all, there is a fall of one hundred and sixty feet. Simple as locks are, I could not help reverting to the wild rapids at Trenton Falls, and reflecting upon how the ingenuity of man had so easily been able to overcome and control Nature!

"The locks did not detain us long - they never lose time in America. When the boat had entered the lock, and the gate was closed upon her, the water was let off with a rapidity which considerably affected her level, and her bows pointed downwards. I timed one lock with a fall of fifteen feet. From the time the gate was closed behind us until the lower one was opened for our egress, was exactly one minute and a quarter; and the boat sank down in the lock so rapidly as to give you the idea that she was scuttled and sinking . . . .

"I was pleased with the journey, although, what with ducking to bridges, bites from mosquitoes, and the constant blowing of their unearthly horn with only one note, and which one must have been borrowed from the gamut of the infernal regions, I had had enough of it."

Lock Number Three on the Oswego Canal at Fulton, with a lift of 27 feet.

THE CHAMPLAIN CANAL AND THE "JUNCTION LOCK"

By Bruce J. Russell

New York State's "North Country" holds enormous fascination for the canal and inland waterway enthusiast as well as for the general historian, combining the still functioning NY State Barge Canal system with the extant remains of 19th Century towpath canals. The scenery is breathtaking, and the region is rich in history, with numerous reminders of the French and Indian and Revolutionary Wars as well as the early period of industrialization when primitive forges, factories, and mills powered by water wheels began dotting the landscape. In 1988 the 190 miles between Albany and the Canadian border continue to attract thousands of vacationers each summer, with the upper Hudson River, Lake George and Lake Champlain perhaps being the major points. However in the years following the War of 1812 and the establishment of peaceful relations with Canada, the most pressing need was transportation. The North Hudson River valley, bordered on the east by the Taconic Hills and on the west by the Adirondack Mountains, contained rich timber and mineral resources, not to mention acres of prime farmland waiting to be cultivated. The northern counties, in spite of their harsh winters, were beginning to attract settlers in significant numbers, and places such as Fort Edward and Whitehall which had previously served as frontier garrisons and trading posts began acquiring substantial population. The need for cheap as well as dependable transportation links with Albany to the south and the St. Lawrence River Valley to the north became more pressing each year. Primitive wagon roads, often impassable in the rainy season, were inadequate to do the job.

Typical dam and lock on the Champlain Canal as it appears today on the canalized section of the Hudson River. (Photo by Bill Shank.)

In 1817 the New York State legislature under the leadership of Gov. DeWitt Clinton authorized the building of two artificial canals to meet this need. The canal best known to students of American history was, of course, the Erie, which opened for business in 1825 amidst much fanfare. Conceived and built a few year prior to the era of the railroads, it provided a through water route between Albany and Buffalo, uniting the Hudson River with the Great Lakes. However in order to obtain enough statewide support for financing and building the original Erie, somewhat irreverently called "Clinton's Ditch", promises for similar internal improvements had to be made to the counties of the North Country. Otherwise their votes could not be counted on for the digging of the principle route to Buffalo. Then, as now, politics was the name of the game.

On account of these developments and the political posturing which followed, a second canal was also approved and the requisite funds allotted to it. This man made waterway was to be known as the Champlain Canal, and its route involved a passage from Fort Edward, the practical limit of navigation on the upper Hudson River, to Whitehall on the southern extremity of Lake Champlain. Here it was possible to sail a boat to the northern end of that 50 mile long body of water. In addition Canadian interests were contemplating a canal from the long narrow lake northward to Quebec province and on to Montreal via the St. Lawrence River. The American and Canadian canals would therefore permit passage of vessels from the Hudson River to the St. Lawrence, or from Albany to Montreal.

Interior of the Mechanicville, New York Lock on the present Champlain Canal, as seen from the deck of the Emita II. (Photo by Bill Shank.)

Original Champlain Canal

An initial survey for the Champlain waterway was undertaken by James Geddes in autumn of 1817, and the basic route from Fort Edward to Whitehall was staked out. Worked commenced almost immediately using gangs of immigrant laborers wielding pick axes and shovels, and by 1818 over 12 miles of ditch had been dug with dimensions conforming to New York's other canal - the Erie. By 1819 the entire stretch from Fort Edward to Whitehall, a distance of about 25 miles, was complete. However the original scheme, perhaps to economize on costs, was for boats to utilize the upper Hudson River from Fort Edward as far south as Saratoga Falls. At this point another man made channel would run as far south as Waterford where it would empty into the Erie. The combined canals would then proceed to Albany where cargos might be transferred to large river vessels, or groups of barges could be tied together and towed by steam tug to New York City.

The southern segment was finished in 1822, three years after the northern part. Nevertheless it appeared illogical to many people to have a situation whereby two artificial, man made canals

(Continued on Page 26)

THE CHAMPLAIN CANAL (Cont.)

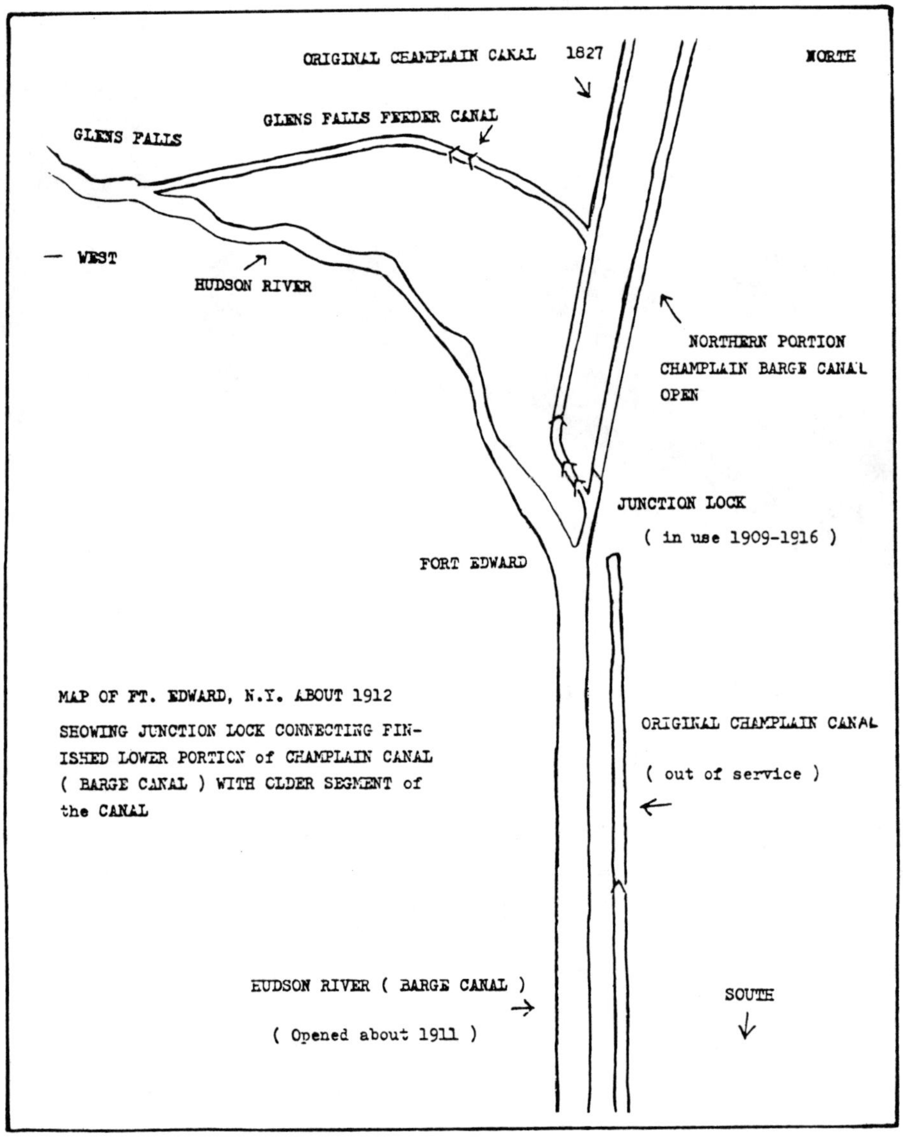

(Continued from Page 25)

were separated by an intermediate section of Hudson River, forcing boats to exit the canal at Fort Edward, travel down the swift flowing river to Saratoga Falls, and then re-enter another canal for the remainder of the journey. Use of the unimproved Hudson for this middle portion created difficulties for the boatmen, and in 1825 it was decided to authorize the digging of a third stretch of artificial waterway so that a continuous canal from Whitehall to Waterford might be formed, with the natural river with its rock filled bottom bypassed entirely. The final section of the Champlain or Northern Canal was finished in 1827. Also in 1829 a short branch or feeder canal was dug from the main channel at Fort Edward to Glens Falls, a distance of about 5 miles. Its purpose was to bring water from another portion of the Hudson River, farther north, into the canal's summit level. Furthermore the Glens Falls feeder was made navigable, since that community constituted a valuable source of traffic.

The dimensions of the original 1827 Champlain Canal were 40 feet wide and 4 feet deep, with stone lock chambers measuring 90 feet by 15 feet. In subsequent years these figures were gradually increased, and by 1877 all locks had been rebuilt to 110 feet in length by 18 feet wide. Additionally a uniform depth was created by removing more material from the bottom of the prism. The enlargement of the Champlain was undertaken for two reasons - one to conform with the Erie Canal, which was simultaneously being improved, and the other to enable bigger, deeper bottomed boats with greater carrying capacity to use it. In 1896 a plan was drafted to expand even further the dimensions of the Champlain and Erie Canals which would have provided for a 7 foot depth and even larger locks. However by this late date in America's Canal Age competitive forces in the form of railroads and surfaced roads were at work, and this final improvement to the New York State inland waterway network was never begun. If water transport was to coexist with the Iron Horse, an all weather form of transportation capable of fast speeds and boats of much greater carrying capacity were needed so that bulk products such as board lumber and coal could be carried in sufficient quantity that a reasonable profit might be generated.

Best Years, 1830-1875

The best years in terms of volume of business and overall profitability for the original Champlain Canal were between 1830 and 1875. Much lumber from the Canadian wilderness was transported via this water route. Its journey often began on the shores of the St. Lawrence River. Subsequently it entered the Chambly Canal which gave access to Lake Champlain, and on to Whitehall. General merchandise including coal was also carried in the mule drawn, wooden boats, providing a lifeline to such North Country communities as Mechanicville, Saratoga, (site of a famous Revolutionary War battle), Fort Edward, Fort Ann, and Glens Falls, not to mention the tiny hamlets situated on the shores of the lake itself. Brutally cold winters meant that the canal had to be shut down from early December to late March, but during the brief summer season traffic along the 56 mile long man made canal was brisk and only began to be eclipsed when nearby railroads such as the Delaware & Hudson came into their own.

Barge Canal System Planned

New York State's decision to upgrade four of its 19th Century towpath canals (the Erie, Oswego, Cayoga-Seneca, and Champlain) into modern 20th Century waterways using a combination of improved rivers and wide, machine excavated channels was made in 1898, and in 1901 surveying commenced. The 1829 Champlain Canal was slated for replacement by a new navigational system involving the upper Hudson River from Waterford to Fort Edward, and a new cut-channel from Fort Edward to Whitehall, complete with electro-mechanically operated locks, large concrete chambers, and wide dimensions sufficient to permit passage of vessels in excess of 250 feet. No provision was made for a towpath; the barges would be self propelled or pulled by steam powered tugboats. The new route of the Champlain Canal would of course parallel the 19th Century waterway from Waterford north to Whitehall and the lake. In 1905 digging began, and the days of the mule-drawn boats in the North Country were numbered.

The work was divided into two principal sections. The southern portion would run from Waterford to Fort Edward and primarily consist of the Hudson River improved for navigation in the same fashion that the Mohawk was upgraded on the Erie Canal. The northern part would consist of the portion from Fort Edward to Whitehall, and would be a man-made channel. Since the southern section or river navigation was being equipped with locks identical to those in the north, it was likewise considered to be a canal. The water flowing through it would be regulated by a series of slack-water navigations.

Work on the two dissimilar sections proceeded steadily. By 1908 it appeared that the lower part, involving the aforementioned improvements to the Hudson River, would be finished long before the upper, which involved the digging of a new 25-mile waterway of wide width. Thus a situation developed in which 50% of the new Champlain Canal was ready for use while the remainder wasn't. Instead the old waterway from Whitehall to Fort Edward with its 19th Century locks had to remain in operation for several more years. In 1909 the southern section (of 1827 vintage) saw most of its traffic diverted into the canalized Hudson River although actual abandonment didn't occur until 1916. Nevertheless in order to maintain through navigation from the Albany/Waterford area to Canada the original "cut" still had to be utilized from Fort Edward to Whitehall at Lake Champlain. Furthermore it would be at least three more years before the northern portion of the new canal came into active usage. Consequently the logical solution was to make a physical connection or junction between the completed new waterway and the existing old canal.

All that remains of the Fort Edward Junction Lock which once connected the completed portion of the Champlain Barge Canal with the older northern segment. Lock was built about 1909 and abandoned approximately 1916 when the Champlain Barge Canal was completed in its entirety. (Bruce Russell photo)

The Junction Lock

Owing to the difference in elevation of several feet between the 19th Century man made Champlain Canal at Fort Edward and the canalized river running parallel to it the connection between the two was designed in the form of a lock. Boats exiting the Hudson would enter it, be raised to the level of the old waterway, and then proceed northward. Of course this would be a temporary situation, to preserve the continuity of inland navigation for a few years in order that work on the northern segment might be completed without interference. Once this was accomplished, the junction lock as well as the remainder of the old Champlain Canal north of Fort Edward could be abandoned and everything routed into the 20th Century barge canal.

In retrospect it might have been simpler to have suspended navigation on the Champlain route until the entire job was finished. However there were at the turn of the century a significant number of commercial customers including lumber mills and coal yards which required service, and if their shipments were interrupted for more than one season they might turn to the competing railroads or perhaps even relocate away from the canal. Hence it was in everyone's best interest to build the junction lock and keep traffic flowing over the whole length of the Champlain Canal.

Construction of this piece of waterway apparatus involved use of hydraulic concrete rather than traditional stone masonry. By 1900 concrete had been perfected and was also being used as the principal material for the 300 foot long locks of the replacement canal. Therefore it was designed as a modern lock except that the actual lock gates would continue to be opened and closed by hand. Also contributing to its uniqueness was the fact that at its south end was a traditional set of wooden miter gates fitted with balance beams, while at its north end existed a steel drop gate equipped with a cranking mechanism using a chain and wheel. A close inspection of the ruins of the chamber revealed recesses in its side walls where the heavy metal gate was raised from the floor by means of the aforementioned crank. Such hybrid lock chambers involving traditional miter gates on one end and a drop gate on the other were also used on other canals such as the Lehigh and the Delaware.

The Junction Lock had a very brief existence - no more than three or four years - since once the new channel came into usage about 1916, all traffic bypassed it. Soon water was drained from the original Champlain and the part through the village of Fort Edward filled in and converted into a road. However the lock itself, perhaps since it was constructed of concrete, was allowed to remain IN SITU, minus its metal drop gate and associated winding mechanism, which were probably removed for use elsewhere. Over the past 70 years it has held up reasonably well, and has been visited on several occasions by various canal societies, the most recent one being in the Spring of 1988.

Members of the Canal Society of New Jersey, under the able direction of Bill Moss, Capt. Bill McKelvey, and Lance Metz organized a two-day weekend tour of the Champlain Canal. One of the highlights was a stop at the Junction Lock at Fort Edward. Unless one uses his or her imagination it is difficult to imagine canal boats passing between its sturdy six foot thick concrete walls, now beginning to crumble. Yet this is precisely what they did for five years. The Junction Lock is perhaps but a minor footnote in the saga of American canaling, but nonetheless an interesting one.

Another view of the Junction Lock at Ford Edward, N.Y. with members of the Canal Society of New Jersey inspecting it. The thick concrete walls haven't seen water for over 70 years, yet they survive as reminders of another era. (Bruce Russell photo)

The last surviving-intact Erie Canal aqueduct may become a shopping mall, an entertainment center, a museum, or simply a pedestrian viaduct, under plans being considered for the revitalization of downtown Rochester. After it ceased to carry the canal over the Genesee River, it was used for a time to carry subway tracks, and, after the subway went out of business, freight train tracks, but it has been idle and closed to the public since the 1960s. We have no date for the article by Sherrie Negea in the Rochester, N.Y. Democrat and Chronicle. (Contributed by Keith Kroon)

WHEN THE IRON HORSE WAS FOALED

An exact replica of the "Stourbridge Lion" as built for the 1933 Century of Progress Exposition in Chicago.

Robert J. Deemer, J.D.

In the early 1820's, a group of New York entrepreneurs under the leadership of two brothers, William and Maurice Wurts, studied the possibility of building a canal from the Hudson River westwardly toward the coal fields of northeastern Pennsylvania. The coal found there was anthracite, or "hard" coal. The canal project was begun in 1823 under the management of the Delaware & Hudson Canal Company which was chartered by the State of New York and the Commonwealth of Pennsylvania.

The route planned for the canal was 108 miles in length extending from Rondout, near Kingston, New York, across the relatively flat portion of southern New York, across the Delaware River and into Pennsylvania, crossing a smaller river, the Lackawaxen, thence to its headwaters at a village called "Dyberry Forks".

By the route planned the canal would terminate about ten miles from the foot of the range of hills known as the "Moosic Mountains".

The coal was being mined on the westerly side of the Moosics near Carbondale. Originally the Company planned to have the coal hauled over the mountain by wagons, deposited at the head of the canal where it could be loaded on canal boats to begin the journey by water to Rondout where the boats would either be unloaded and the coal trans-shipped down the Hudson, or the boats themselves could be towed by steam tugs to New York for unloading.

Construction of the canal began in 1825 when Philip Hone, then Mayor of New York and first president of the Company, turned the first shovelful of dirt. The work moved quickly and on October 20, 1828, the canal was considered completed.

A pleasant bit of history had its origin when Mayor Hone, in company with other notables including the famous American author, Washington Irving, made an inspection trip of the canal. Upon reaching the village of Dyberry Forks, it was suggested that the place be re-named "Honesdale" to honor the leader of the great achievement. Not to be outdone in courtesy due the eminent Mr. Irving, observing a high craggy eminence on the easterly side of the town, it was suggested that the rocky cliff should be named "Irving Cliff". And so both names remain to this day.

The Canal Company had for its chief engineer, John B. Jervis, a civil engineer who earned a high reputation when he was active in the construction of the Erie Canal. It is he for whom the New York railroad town "Port Jervis" is named.

Mr. Jervis had with him a young associate who showed considerable promise as an engineer, a graduate of Columbia College, New York, in 1823. His name was Horatio Allen.

Gravity Railroad

Although the Canal Company had under construction a form of "gravity" railroad to be used to bring coal over the Moosics, the officials thought there might be a better way to get the result desired—dependable, economical and efficient means of moving large quantities of coal loaded in "wagons".

Being aware of development in England, in 1828 the Company appointed Mr. Allen to go to England to study the possibility of using steam as a motive power for land vehicles. It had been used successfully for propulsion of vessels on water but steam had not yet been adapted to use in moving vehicles on land.

Mr. Allen had authority to purchase four engines in England as well as a quantity of iron rolled for use in the construction of tracks for such engines.

Allen's mission was begun by his purchase of three engines from the firm of Messrs. Foster, Rastrick & Co., all of a type then in manufacture in England. He still had unused authority for one more engine—but this was destined to be the most famous of all of them.

In those days an English engineer, George Stephenson, was preeminent among the experimenters in the field of stream locomotion. Fortunately, Mr. Allen met with Stephenson and gained from him considerable advice and instruction based on Stephenson's experience which had begun in 1814.

Mr. Allen placed the order for the remaining locomotive with Stephenson & Company. It was intended that the engine would be built along the same general plan as Stephenson's engine, "Rocket" which was to gain considerable fame in England.

Although originally intended to weigh between four and five tons the engine actually weighed closer to seven tons when first set upon the rails. The distance between the wheels was four feet three inches, not far from what later became the standard gauge for American railroads. As planned, it was expected that the engine would be able to move at a rate of four miles an hour.

"The Lion"

Although there is some doubt as to the final identity and the ultimate disposition of the other three locomotives purchased by Mr. Allen, there is no doubt as to the identity of the fourth engine built by Stephenson & Co. The engine was built in the English town of Stourbridge and during its construction an imaginative painter saw fit to paint the head of a lion on the convex head of the boiler, thus providing the final touch to name the engine "Stourbridge Lion".

The Stourbridge Lion left Liverpool in April, 1829 and arrived at New York on May 13 of that year. Delivered, the engine cost the Company $2,914.90. Upon being unloaded from the ship "John Jay", it was taken to the shops of William Kemble at his West Point Foundry, where it was

Horatio Allen, who brought the "Stourbridge Lion" from England.

assembled, set up on blocks and demonstrated there under steam.

The Lion was then sent up the Hudson to Rondout and then, from Eddyville, it began its trip by canal boat to Honesdale, on July 16, 1829. On August 5, Mr. Jervis reported that the Lion was placed on the railroad and ready for steam.

The route was planned and specially built. It began with about six hundred feet parallel with the canal at Honesdale then it crossed the Lackawaxen Creek on a trestle thirty feet above the water, then on a gentle curve nearly a quarter of a mile long. The curve had a radius of seven hundred and fifty feet. At the end of the curve the road continued in nearly a straight line into the wooded area near what was later called "Seeleyville" a small community along U.S.-Pa. Route 6, some three miles west of Honesdale.

The track itself was made of hemlock stringers, six inches by twelve and set on edge, each stringer being twenty to thirty feet long. The stringers, in turn, were held an equal distance apart by cross ties about ten to fifteen feet distant from each other. The cross ties were supported by posts set in broken stone or on stone piers.

The running surface of the rails was topped by wrought iron straps two and a half inches wide, a half inch thick and about fifteen and a half feet long.

With men waving their hats and ladies staring in astonishment while youths and small boys aided in the production of general noise and confusion, Mr. Allen took off from the place of the engine near the canal. Fortunately for posterity, Mr. Allen left us a description of his trip:

Allen's Trip

"When the steam was of right pressure, and all was ready, I took my position on the platform of the locomotive alone, and with my hand on the throttle-valve handle, said, 'If there is any danger in this ride, it is not necessary that the life and limbs of more than one should be submitted to the danger' and felt that the time would come when I should look back with great interest to the ride then before me.

"The locomotive having no train behind it answered at once to the movement of the valve; soon the straight line was run over, the curve was reached and passed before there was time to think as to its being passed safely, and soon I was out of sight in the three miles' ride alone in the woods of Pennsylvania.

"I had never run a locomotive nor any other engine before. I have never run once since, but on August 9th (sic), I ran that locomotive three miles and back to the place of beginning, and being without experience and without a brakeman, I stopped the locomotive on its return at the place of starting. After losing the cheers of the lookers-on, the only sound, in addition to that of the exhaust steam, was that of the creaking of the timber structure."

This was the first operation of a railroad locomotive in America and the Western Hemisphere and it was two months prior to the Rainhill trials on the Liverpool and Manchester railroad in England where the Stephenson Rocket established there the practicability of steam railroad transportation.

After careful study, it was considered that the engine was too heavy for practical use on this sort of track. As an additional test, however, on September 9, 1829, there was another trial. We have Mr. Allen's report of this event:

"On Wednesday last we had the engine in motion again and its operation and effect on the road carefully observed. The railroad as it now stands is not sufficiently stable for the operation of the Locomotive. Before it is put to work on it the road ought to be carefully examined and strengthened, without doing so it would be unsafe to put the engine to work."

Following the second trial the engine was taken from the rails and stored near the track. The Stourbridge Lion appeared to have been discarded after making its contribution to history. The engine was given some roughcovering until about 1849 when it was taken to Carbondale, Pa., where the boiler was put to use in the shops of the D. & H. Company and some parts worked up for other use. The boiler was used for about twenty years and then replaced by another source of power. Later the boiler was sold and in the 1880's it was found again at Carbondale. On June 18, 1889, the boiler was deposited in the Smithsonian Institution in Washington, D.C. A few other parts were found and the engine partially restored and reconstructed now stands in the Smithsonian.

A replica, was built and is housed at Honesdale, Pa., close to the place where the Lion achieved fame on its original run.

Although the Lion had only a brief moment of glory, it was followed quickly by others who imitated it in the use of steam as the source of power for land transportation.

Horatio Allen, who operated the Lion during its brief period of fame, moved on to a distinguished career as an engineer. He was a consulting engineer for the New York & Erie Railroad and later President of the Erie Railroad. Mr. Allen was a consulting engineer in the construction of the famed Brooklyn Bridge and the Panama Railroad, as well as the Croton Aqueduct that supplied water to New York City. After a life full of accomplishments and enjoying the recognition of those of his calling, Mr. Allen died in 1890.

But Mr. Allen was not forgotten by his early employer, the Delaware & Hudson Canal Company and its successor, the Delaware & Hudson Railroad: on December 4, 1924, the D. & H. Company's big, new locomotive #1400, was christened at Colonie, NY., as "The Horatio Allen" in honor of Mr. Allen's services in connection with the celebrated Lion.

While there were many successors reaching the American scene quickly after the Lion, it must be conceded that the Iron Horse was indeed foaled at Honesdale, Pa., August 8, 1829.

GENESEE VALLEY CANAL — 150TH

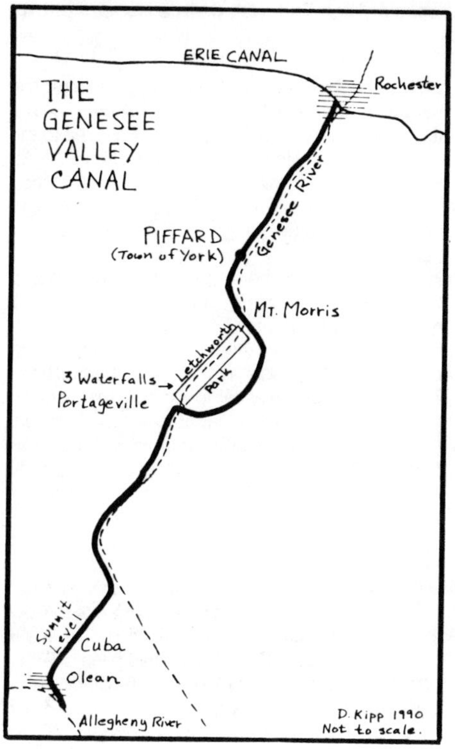

By David L. Kipp

On September 22, 1990, the Historic Society of the town of York, NY sponsored a festival to commemorate the 150th anniversary of the September 1, 1840 opening of the Genesee Valley Canal (GVC) from Rochester to Mt. Morris. The GVC was later completed to Olean, NY connecting with the Allegheny River. The celebration took place in the hamlet of Piffard which straddles the old canal line about 3 miles west of Geneseo, NY. Events took place in and around a canal era inn recently and beautifully restored as the "Yard of Ale" restaurant.

The Canal Society of NY staffed a booth which drew many inquirers to its displays of maps, booklets, site photographs and slides relating to the GVC and other NY canals. All the available guidebooks for canal tours published by the Society were sold out within the 1st 2 hours.

The highlight of the day was the dedication of an historic marker inscribed as follows:

"GENESEE VALLEY CANAL Rochester to Olean 1836-1878. This section of the Genesee Valley Canal opened as far as Mt. Morris on September 1, 1840. The canal provided a practical means of transporting both goods and people. making Piffard an active commercial center. TOWN OF YORK HISTORICAL SOCIETY September 22, 1990."

The dedication ceremony was heralded by blasts on an authentic GVC boatman's horn provided by Mrs. Emily Madden of Livonia, NY who, with her late husband Heywood, accumulated several decade's worth of field trip notes, photographs and research to become THE experts on GVC history and lore. It was punctuated by a musket salute from Little Beard's Men Frontier Encampment at the moment of the unveiling.

Among those lending their words to the moment were Tom Grasso, President of the Canal Society of NY. He pointed out that although not an economic success, the GVC should be remembered with pride for its role in the settlement and development of the Genesee Valley and for the unique engineering challenges it overcame. The canal climbed 978 feet from its Rochester junction with the Erie Canal to the summit level near Cuba, NY, having 50 locks in one 5 mile stretch as it by-passed the 3 waterfalls of the Letchworth Park Canyon. Congresswomen Louise Slaughter praised the local citizenry for celebrating their unique heritage in a way that will enhance the attractiveness and historical interest of the Valley. Bud Van Arsdale, who is on the board of directors of the NY Parks and Conservation Association which is attempting to develop a greenway linear park along the Genesee River corridor from Rochester to Letchworth Park, explained that the GVC is a vital link in tying the greenway concept together.

This event hopefully portends an awakening awareness of this little known canal and looks to the possibility of projects that will preserve and highlight what remains of the GVC. In focusing on such monuments to our exciting transportation, engineering and economic past, we can make them more accessible and accurately interpreted to the general public.

CANAL SOCIETY OF NEW JERSEY VISITS THE D&H

This shows the excellent condition of many of the surviving locks of the D. & H. Canal, nearly a century after its abandonment, One of the locks at High Fall, NY. (Bruce Russell photo)

By Bruce Russell
Contributing Editor

THE CANAL SOCIETY of NEW JERSEY has done it again! For the 11th year in a row they have planned, organized, and executed an exploratory trip over an existing or abandoned canal PAR EXCELLENCE. Advance preparation and preliminary scouting of the territory by BOB BARTH, LINDA HOUSE, and BILL MOSS combined to insure that everything went according to plan. New Jersey's "canalers" were certainly ready for a trip after five months of winter "hibernation", and two bus loads of men and women attested to their enthusiasm.

It would be impossible to fully enjoy a trip such as this without knowing something of the history of the DELAWARE and HUDSON CANAL. Thoughtfully BILL MOSS had included with everyone's information packet a fact sheet giving key dates pertaining to the waterway, which opened in 1828 and was abandoned except for a short stretch near Kingston in 1898, after the railroads had supplanted it as a mover of anthracite coal from the Carbondale, Pa. area to the major market of New York City. The D&H CANAL was actually the middle link in a transportation system that began near Carbondale (close to Scranton) with a unique gravity railroad which carried coal up and over Moosic Mountain to Honesdale, Pa. where the waterway actually began. After traversing the approximately 108 miles of canal in boats of 140 ton capacity pulled by mule teams, the fuel was unloaded at the mouth of the Rondout Creek, a tributary of the Hudson. The place where it was removed from the wooden vessels was known in canal days as Terminal Island, a man-made facility where it was stored in gigantic piles. At the end of summer or during early autumn it was reloaded into much larger barges which were then lashed together to form a "tow". This was taken down the Hudson by steam tug to New York City.

At Rosendale this "Canal siding" ran off the main channel directly into the Snyder cement works to allow canal boats to unload coal for the kilns and pick up bags of newly-made cement. (Russell photo)

On the second day of our trip there was a boat ride which gave tour participants a chance to view the remains of Terminal Island which still sits in the middle of Rondout Creek, minus its piles of D&H coal and the machinery used to transfer it from canal barge to river barge.

MAURICE and WILLIAM WURTZ of Philadelphia initially proposed the idea of an artificial waterway or canal connecting the Pennsylvania anthracite fields with the Hudson River in the years immediately following the War of 1812. Wood as a fuel was no longer available in sufficient quantities to supply a growing nation and anthracite coal, first discovered in the 1790s, was the answer. It burned both hot and relatively clean. They hired JOHN JERVIS and BENJAMIN WRIGHT who had perfected their canal building skills on the ERIE, a public project as opposed to the D&H which was a private venture. These two engineers realized that a canal traveling in a straight line connecting Honesdale with Kingston on the Hudson was impossible due to the Shawangunk Mountains, so they designed a waterway which would initially head south by following the valley of the Lackawaxen and then the Delaware River. At Port Jervis, N.Y. it would proceed north towards the Hudson by paralleling the Neversink and Rondout Creeks from which it would also draw water. Hence a map of the D&H CANAL resembles a giant V with Port Jervis at the bottom. Our field trip began at Port Jervis and concentrated on the remains of the waterway as far north as Rondout near Kingston. On previous trips we had explored the Honesdale to Port Jervis section as well as the previously mentioned gravity lines.

Ground breaking for the D&H CANAL occurred on July 13, 1825 at Beatysburg, N.Y. which is now appropriately called Wurtsboro. On our April 28 and 29th trip we stopped briefly at the point where the Wurtz Brothers turned the first shovelful of earth to initiate work on the waterway, which was finished three years later in 1828 at a cost of $1,200,000. A plaque is affixed to an old stone snubbing post marking the spot where it all began. Also at this location was a storage basin and boat building facility.

Locally quarried sandstone was utilized to fabricate the lock chambers and aqueducts, and by sheer accident a product known as natural cement was discovered in the area of Rosendale during the construction period. With its ability to harden and become super strong under water it was an ideal material for building canal structures. Consequently the D&H CANAL, in addition to coal, transported large quantities of this commodity which was produced by the A.J. SNYDER family at their kilns in and around Rosendale.

Snyder Cement Works

As part of the tour we visited the SNYDER ESTATE and cement works.

It took about five days for a boat to travel from Honesdale to Rondout, just below Kingston. Passenger packets took less time but by 1860 had vanished due to railroad competition. BOB BARTH described the MERCHANT'S and TANNER'S LINE which operated a fleet of 6 passenger vessels along the D&H between Ellenville and Rondout. These boats were pulled by horses rather than mules. Our stop at the D&H CANAL MUSEUM located in a former church in High Falls, N.Y. was valuable since it contained a number of exhibits which depicted the lives of D&H boatmen and their families.

As we drove along the highway parallel to the remaining portions of the canal we were able to

Members of the Canal Society of New Jersey are shown at the entrance to the natural cement mine located on the Snyder Cement property at Rosendale, NY. (Russell photo)

The remaining abutments of John Roebling's Aqueduct over Rondout Creek at High Falls, NY. This was a cable-supported structure similar to those built by Roebling at Pittsburgh and Lackawaxen, but with a single span only. The wooden trough burned in 1918. (Bruce Russell photo)

spot various buildings which had once been canal supply houses and stores. Near Lock 50 BOB BARTH pointed out an old red barn which once sheltered D&H CANAL mules. At its peak in the 1870s there were about 1500 boats operating on the D&H, and this meant a lot of hungry and tired men needing to be fed and resupplied. After the waterway closed in 1898 most of these establishments continued in use serving a non-canal clientele, primarily local farmers. We didn't stop at any of them but since several sat on "Canal Street" or "Canal Avenue" it was obvious what their original purpose was.

In Ellenville we passed one of two surviving D&H CANAL toll houses which had been moved from its original site at the time the canal prism was filled in during the early part of this century. Furthermore near Ellenville was situated the DAVIS HOUSE where, according to local legend, the peanut butter and jelly sandwich was invented. It was sold to D&H boatmen passing through town and was valued as a source of quick energy.

Undeveloped Area

One of the reasons why so much of the D&H CANAL survives is that the area through which it ran has not yet fallen victim to rampant real estate development. Thus at High Falls a 5 lock flight remains intact along a half mile stretch beginning at the previously discussed museum. Unfortunately some of the other locks as well as sections of the dewatered prism are now situated on private property and no longer accessible to the casual tourist. Nevertheless the "Advance Party" for our trip managed to contact some of the people to get permission to enter their premises. There were a few isolated canal segments which still contained water, giving a more authentic, historically accurate picture. Generally these remnants are fed by underground streams and the clay which lined the old canal bottom to prevent seepage continues to retain water, incredible, after 160 years!

At High Falls across the road from the museum were the ruins of two parallel aqueducts which once crossed Rondout Creek. The first of which almost nothing survives was the original 1828 structure consisting of twin masonry arches. It was designed for boats of 10 ton capacity. After 1845 when the waterway was enlarged for the third time to provide a 6 foot depth it became useless since the larger capacity boats couldn't fit through it. Consequently next to it a larger replacement was built. This was not a twin arch affair but instead was a suspension aqueduct designed by JOHN ROEBLING, of Brooklyn Bridge fame. An enormous wooden trough made water tight with bitumin carried the D&H CANAL approximately 40 feet over the river. This trough was supported by iron rods and steel cables, portions of which remain embedded in the surviving stonework.

As part of the 1845 aqueduct a mile long new alignment or approach to High Falls was built. Our N.Y. STATE tour guides explained that for this section there were actually two PARALLEL D&H CANALS. One was the orignal 1828 prism which was not enlarged or deepened, and the new one which replaced it. Each had its own locks and as mentioned above its own aqueduct at Rondout. During our 2 hour stop at High Falls we were able

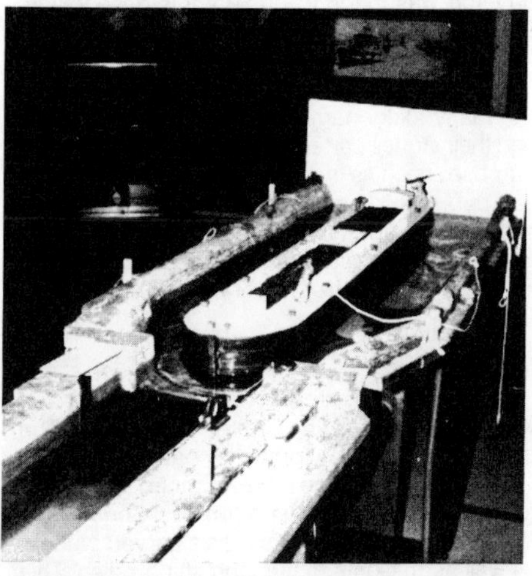

The D. & H. Canal Museum at High Falls contains several interesting models and dioramas showing how the canal locks worked. This is one of them. (Bruce Russell photo)

to explore the remains of both aqueducts and prisms. Walking back into the hills behind the museum we located the two parallel flights of locks - one for each alignment. However this situation was not common. In most instances when the canal was made bigger to handle larger boats the existing channel was widened and deepened and its locks reconstructed to new dimensions. High Falls was one of two exceptions. The museum offered a brochure describing all of this plus directions for a self guided walking tour. It was a most worthwhile stop.

(To be concluded)

LETTER TO THE EDITOR

In Bruce Russell's account of the Canadian-American meeting in issue no. 71, I was interested in his paragraph: "The slide show brought back memories of both Great Lakes shipping and life along the canal, and it's sad to think that in the U.S.A. and Canada there are no longer companies which operate large luxury boats between various ports on the five major lakes and connecting rivers."

Yes, indeed, it *is* sad. But I feel myself fortunate to have enjoyed some of what he recalls in memory.

About 1938, I rode a Hudson River Day Line boat from Albany to New York City on a vessel that had a walking beam to transmit power from the steam engine to the side-wheel cranks. Such boats as may now remain do no more than short excursions on the river.

In late summer 1944, when I was working in Cloquet, Minnesota, west of Duluth, and my parents were in a Chicago suburb, I went home to visit them, then took the "North American" of the Chicago, Duluth & Georgian Bay Transit Co.—"Georgian Bay Line"—from Chicago to Buffalo, lay over a couple of days, then took the "South American" from there to Duluth. It was a good trip through four of the five Great Lakes.

In my copy of the Official Railway Guide dated July 1931, a full-page ad offers "A week's cruise on four lakes, $72.50, including meals and berth." Imagine! Nearby in the Guide is a full-page ad of the Hudson River Day Line, offering the "Hudson River by Daylight."

The Eastern Steamship Lines have two facing pages, and when I was in college in Cambridge, Massachusetts, in the 1930s, I took both their boats, the "Boston" and the "New York," between those two cities; it was an overnight trip through the Cap Cod Canal.

And during that interval, also, I managed to ride the old Fall River Line; the boat train left Boston's South Station and ran right out onto the pier at Fall River for the transfer of but a few steps onto the side-wheel vessel, which took us overnight to New York. I was fortunate in being able to wangle my way onto the bridge as we went down Narragansett Bay.

All those interesting ways of travel are long gone, and it's a great shame. One would think that people today would enjoy a water trip between lake, ocean, and river cities. And that reminds me that I once took, in 1942, a stern-wheel Mississippi River boat, the "Golden Eagle," from St. Louis to Peoria and back—a great trip.

Also, while in Boston I wangled my way onto a tug that went up to Beverly to undock a coal boat; and onto a steam locomotive from Boston to Fitchburg on the Boston & Maine Railroad. And a while ago, you published my account of a trip on an Erie Canal tug.

Gerald Chapman, P.O. Box 464, Stockbridge, Massachusetts, 01262.

NEW JERSEY CANALERS VISIT THE D & H — CONCLUSION

Lock Number 16 on the D. & H. Canal at High Falls, New York. This was one of a series of five locks (Numbers 16-20) that took the canal over a hill. They are part of a re-building of the canal which occurred in 1848-50, by-passing the original right-of-way. (Russell photo.)

By Bruce Russell

On Saturday evening April 28 the group stayed at the Kingston, N.Y. Holiday Inn.

Following an "all you could eat" buffet supper we were treated to an excellent slide show by GAYLE GRUNWALD and DIETRICH WARNER, owners of the SYNDER estate. Using old pictures and other visual aids they explained the process of making natural cement from limestone, and illustrated how it was transported in D&H canalboats. Interestingly a type of "canal siding" veered off from the main channel and went directly into the SNYDER property where coal to fire the kilns was removed and cement put in its place. Natural cement from Rosendale was eventually replaced by Portland type which was considered superior. In 1970 all production at Rosendale ceased. After the D&H CANAL went out of business in 1898 a short section from just north of High Falls to Rondout was maintained for the benefit of the SNYDER cement interests. Thus several of the old and weary locks continued to be emptied and filled each year so that cement laden boats could reach Rondout. A washout in 1905 severely damaged this remnant of the 108 mile long, 110 lock waterway and it wasn't repaired. Apparently by this time a decent road had been built and the cement came out of Rosendale by truck. Thus ended for good the waterway conceived of and built 77 years earlier by MAURICE and WILLIAM WURTZ.

Our first stop on Sunday April 29 was at the site of Tidewater Lock #1 in Eddyville just below Rondout. This was the last lock the boats would pass through prior to reaching TERMINAL ISLAND where they were unloaded. The lock itself plus a piece of its wooden gate survives although it's now situated on private property owned by MR. CHARLES OPITZ, a gentleman in his 70s. After the canal quit in 1905 the land around Lock #1 was acquired by the CENTRAL HUDSON GAS and ELECTRIC COMPANY as part of a real estate investment which never panned out. In the 1930s they decided to build at another location a marina and dock for yachts owned by their top executives. In order to complete this facility as fast as possible and within budget they removed the cap stones around the lock plus two other layers of the heavy blocks. At this time no one was concerned with preserving the remains of a derelict canal and no effort was made to prevent this desecration. So away went the blocks! According to LANCE METZ, canal historian and industrial archeologist in residence at HUGH MOORE PARK in Easton, Pa. many lock chambers of abandoned canals had their stones carted away to be used for other purposes such as the foundations of monuments and public buildings. It has only been during the past two decades that the historic preservation movement has entered the picture and enacted legislation against this type of wanton destruction.

In connection with the D. & H. Canal Tour, Bill Moss asks us to call attention to the extensive planning work that was done by **Paul Ross** and **Barbara Murry** of Ellenville, New York, not to mention their service as guides during the tour.

Weighlock Ruins

At the Lock #1 site we visited the remains of what had been the D&H CANAL's weighlock. The actual chamber now forms the basement of Mr. OPITZ'S home, but he kindly opened it to permit tour participants a first hand glimpse. Some of the stones still had boatmen's initials carved in them as well as mason's marks and lifting holes. This weighlock was similar to the one used by the ERIE CANAL at Syracuse, N.Y. which survives although minus its machinery. Boats entered the weighlock chamber, water was drained until they sat inside a type of wooden cradle. A system of fulcrums, weights and counterweights was employed to somehow arrive at a figure for the vessel's tonnage. Mr. OPITZ explained that during his walks along the old towpath he still finds hunks of coal. His explanation was that the boatmen used to throw pieces of anthracite at rabbits, squirrels, and other animals while waiting to get into either Lock #1 or the adjoining weighlock. While they were usually unsuccessful in striking their intended target it was a means of killing time. He also told of how during the Depression of the 1930s people would board rowboats and canoes and travel to TERMINAL ISLAND where they searched for pieces of coal left over from the days when it was a busy unloading facility.

"Creek Locks"

Besides being the eastern end of the D&H CANAL, Eddyville also contains buildings which once overlooked the waterway. A couple of saloons and a general store still survive but in somewhat altered form. We then moved on to Locks #2 and #3, known as the CREEK LOCKS. Both were in good condition and were examined closely by the group. Then it was on to Rondout which is actually the harbor of Kingston. After visiting the TROLLEY MUSEUM of NEW YORK which operates a 1921 vintage gasoline railcar over a short piece of the old ULSTER and DELAWARE RR we boarded the RONDOUT BELLE, an old fashioned but diesel powered excursion vessel. From it we were able to get a close up look at TERMINAL ISLAND plus the light house which marks the point where the Rondout Creek merges with the Hudson River. In the 1880s and 90s Rondout was a very busy place with D&H canal boats being unloaded, tows of larger barges being assembled for the trip down the Hudson, and boatmen hanging around between tours of

This wooden gate still remains partly intact on Lock No. 1 at Rondout, New York, almost 85 years after abandonment. At this point, the canal boats left the canal and entered Rondout Creek and traveled downstream one mile to the final terminus at Island Dock, for transhipment to New York City. (Russell photo.)

One of two surviving toll houses on the D. & H. Canal. This 150-year old building is located at Ellenville, New York, but has been moved from its original location when the canal bed was filled in for a local street in the 1920's. (Russell photo.)

duty. A picture in COAL BOATS to TIDEWATER shows hundreds of D&H canal boats laid up at TERMINAL ISLAND during the winter. Rondout was also where the great paddle wheel steamer MARY POWELL docked after her run from New York City. It's a quiet place today but efforts are being made to restore the historic district. A small maritime museum has been built and our group visited it. Several models of the enormous steamboats which once graced the majestic Hudson were on display as well as enlarged photographs. The past is gone and can never be brought back, but at least efforts are being made to capture a tiny portion of it.

Upon departure from Rondout harbor our busses made a 1 hour stop at the historic downtown section of Kingston where many old, stately Victorian buildings were being restored to their previous grandeur. Kingston at the turn of the century was a prosperous community, its wealth coming from the D&H Canal, the steamboats, and the West Shore Railroad which still exists but only as a Conrail freight line. A walking tour and closer examination of its stately mansions, churches, and public buildings attests to this now faded prosperity. Hopefully the city will stage a comeback since the mid-Hudson River region where it's situated is breathtakingly beautiful. From the heights one gets a sweeping view of the lordly Hudson, and while winters are cold, summers are refreshingly cool.

After reboarding our CONOVER BROTHERS coaches it was back to New Jersey, another top notch CANAL SOCIETY of NEW JERSEY field trip completed. Heading south along scenic back roads we once more paralleled the abandoned D&H CANAL and a number of its ancient lock chambers could be spotted. Appropriately someone had put a tape of canal era music onto the bus's stereo/public address system. Our two day journey in D&H country was coming to a close, an event to be savored for many years.

ERIE CANAL

The original canal system paid for its self in only a few years of operation. Then for a considerable time it produced a profit for the state. But with the development of rail transportation and the competition it offered, canal traffic began to dwindle. To encourage use of state waterways a State Constitution amendment eliminated all tolls on the canals.

One recommendation now under consideration by those planning the future of the system involves the imposition of fees for passage on the canals.

Recreational vessels make up the bulk of present day canal traffic and this additional tax would most directly effect them. Reaction to this proposal should be monitored by the canal cruising public and decisions regarding its merits forwarded to legislators involved.

(From Canawler BARGE, May 1989)

THE DELAWARE & HUDSON CANAL HERITAGE CORRIDOR PROJECT

Over the past two years, our members have noticed articles in local papers about a group called the D & H Canal Heritage Corridor Project. Since the goals of this project are so close to our own, and since the project will continue to be in the news this summer, we thought you might be interested in a review of its purposes and progress:

The D & H Steering Committee was formed in the summer of 1988 by NYPCA — The New York Parks and Conservation Association, a non-profit environmental group working for the recognition, promotion, and protection of state parks, open spaces and greenways. The D & H Steering Committee, made up of local residents (elected officials, business and civic leaders), hopes to preserve and revitalize the D & H Canal Corridor in Ulster County and explore its potential for recreational use. The D & H Canal Historical Society has been represented on this steering committee by Trustee Gayle Grunwald since its founding; Ron Schade of New Paltz is currently serving as Steering Committee Chairman.

After a couple of years of planning, the Corridor project is really taking off. With the assistance of NYPCA, the towns of Rochester, Marbletown and Hurley have received approximately $300,00 in state matching funds to develop municipal linear parks (along the old O & W route in Hurley and the canal bed in the Town of Rochester). NYPCA has also hired Vicki Schroth, a graduate student in Public History at SUNY Albany, to coordinate local efforts for preservation and revitalization of the canal corridor. Vicki will be working out of the Canal Museum this season, and we are delighted to be able to provide a home for this exciting endeavor. Any members interested in participating in this project should call or write the museum to be added to the mailing list.

(From the D. & H. Canal Historial Society Newsletter *for April 1990.)*

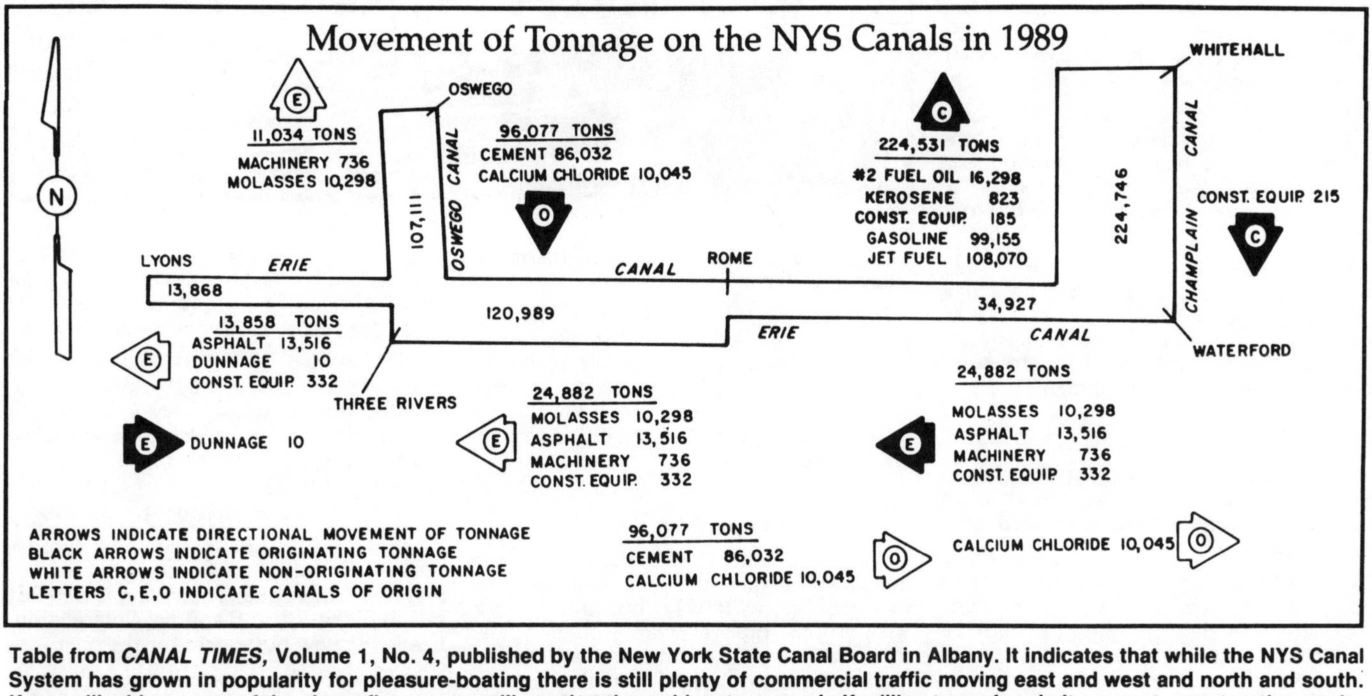

Table from *CANAL TIMES*, Volume 1, No. 4, published by the New York State Canal Board in Albany. It indicates that while the NYS Canal System has grown in popularity for pleasure-boating there is still plenty of commercial traffic moving east and west and north and south. If you will add up some of the above figures you will see that they add up to over a half-million tons of asphalt, cement, construction equipment, fuel oil, molasses, etc. which were shipped on the canals in 1989. (Courtesy of Walter Meseck)

GOWANUS CANAL FUTURE IS "MURKY"!!

By CHARLES SEATON
Daily News Staff Writer

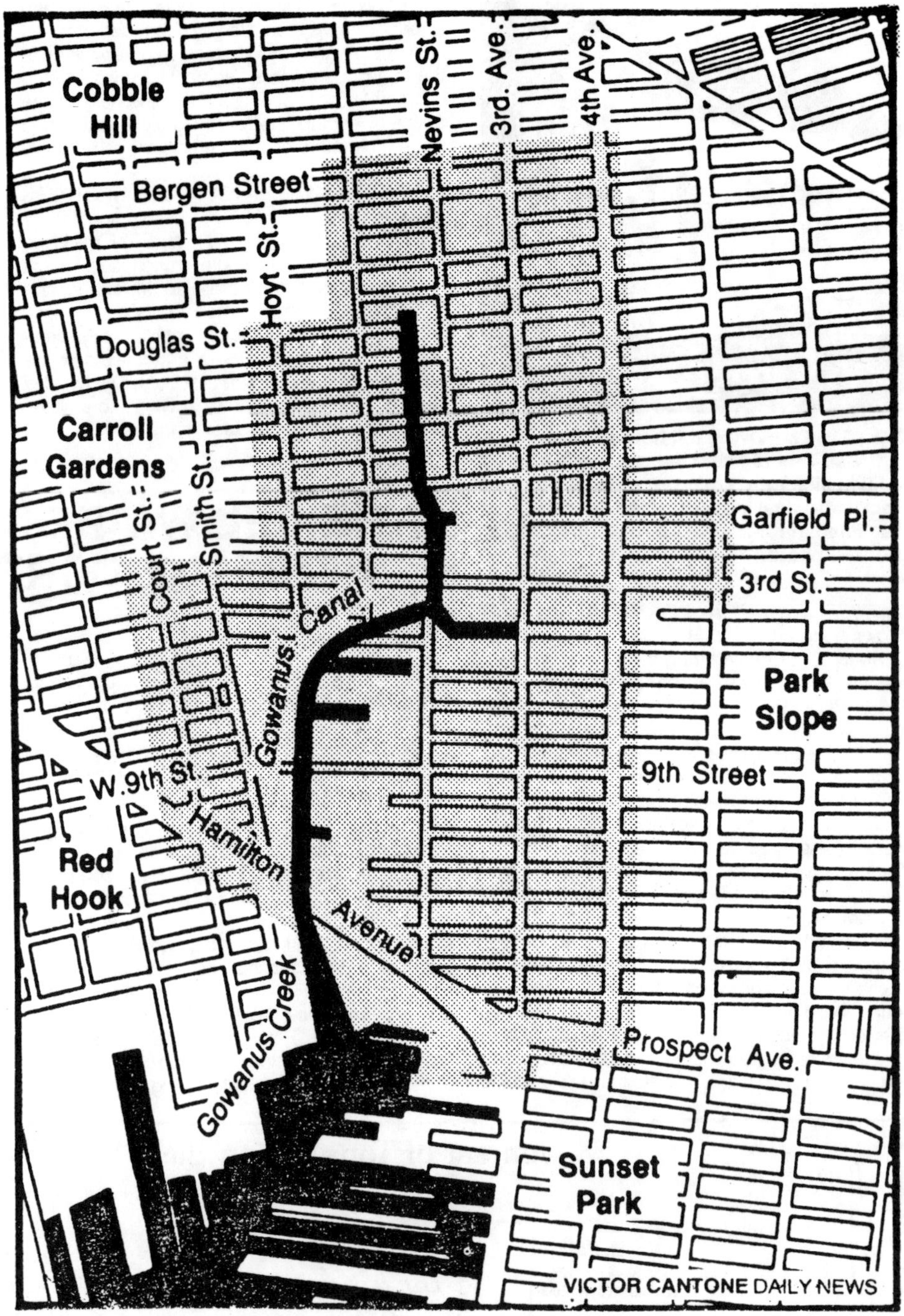

VICTOR CANTONE DAILY NEWS

Although now viewed as an eyesore at best and a health hazard at worst, Brooklyn's much maligned Gowanus Canal was for nearly a century the center of the borough's waterborne trade.

The Gowanus Expressway, however, which diverted much of the waterway traffic from boats to 18-wheeler trucks, was blamed as one of the reasons for the decline in the canal's use and maintenance.

Although a 15-year rebuilding program for the canal is to begin in two weeks, its future — and that of surrounding areas — has been made unclear by controversy.

One segment is asking that the areas around the canal remain industrial, while another is pressing for a combination of business and much needed housing.

Use declined

From the late 1800s until the end of World War II, the Gowanus Canal was the center of a bustling, thriving, manufacturing zone. The 100-foot-wide man-made waterway was created to transport raw materials and finished goods in an age before tractor-trailers, expressways and interstate highway systems.

But use of the canal declined dramatically with construction of the Gowanus Expressway, which freed many of the canal-based firms to trade their barges for trucks.

With its decline came a subsequent lack of maintenance. A pumping station, built in 1947 to keep the murky waters moving, was kept operational for 13 years.

A chief's name

Viewed on a map, the canal appears as a nasty slash running from Gowanus Bay into the heart of Boerum Hill. In reality, the mile-long waterway has earned a well-deserved reputation of being one of the dirtiest in the city.

When the temperature climbs, the stagnant waters bear a palpable odor as far as three blocks from its banks, of which residents have complained — long and often.

The project to construct the canal, named after a Mohawk Indian chief, Gowane, began in 1848. The goal was to rid southern Brooklyn of several acres of marshland while opening the surrounding area to commercial development.

Cost was 88G

Initial cost of the project was nearly $80,000, plus an additional expense of $8,000 for the construction of a boat basin on the bay.

Conditions began to go downhill in 1888 when a bill was passed in the State Legislature to allow the City of Brooklyn to channel rainwater and sewer runoff into the canal at Butler St. Because of the odors spawned by the raw sewage, neighborhood residents dubbed the canal "Lavender Lake."

The waterway's deterioration can be blamed on inadequate maintenance and its use as literally an open air sewer. A study issued in the late 1970s by the Department of Environmental Protection labeled the canal one of the city's dirtiest waterways.

Yuppies nearby

Businesses and their storage facilities are tucked behind the bulkheads on both banks of the waterway, but travel more than a few blocks away from the immediate area and one finds the burgeoning Yuppie communities of Carroll Gardens, Boerum Hill, Park Slope and Sunset Park.

As the massive clean-up and rehabilitation campaign begins, however, sides are already being chosen to determine just what the future of the canal area will be.

The choices are either an increased emphasis on manufacturing and industry, or the beginning of a new residential community in the midst of a group of growing Brooklyn communities.

Business sought

According to a recent city study, businesses which remain along the waterway are still strong. The study, issued last fall by the City Planning Commission, is seeking new businesses for the area.

To protect the industrial district around the canal, the study proposed the rezoning of several areas for light manufacturing to encourage the retention and expansion of industrial jobs. Currently, the areas are zoned residential.

Another strategy seeks the prevention of illegal residential conversion along the canal through monitoring by the Loft Board. The study also recommends inducing firms moving out of Manhattan to select sites near the canal through grants from the newly established Industrial Relocation and Retention Program.

Development pushed

Salvatore Scotto, one of the founders of the Gowanus Canal Development Corporation and president of the Carroll Gardens Association, takes a different view, one that would push the development of residential properties on the west bank of the canal and the continued development of light industry on the opposite side. "We view the canal as an incredible resource," said Scotto. "It is a great drawing point for low-to-moderate income housing and a light manufacturing district."

As an example of the rebirth of the area, Scotto cited the conversion of a 10-story factory building into a 126-unit cooperative at 505 Court St. The conversion spurred the rehabilitation of four adjacent buildings and has led to a stabilization of the area.

Housing needed

Scotto added that low-income and senior citizen housing is desperately needed in the area and if the canal was to remain, it should be developed to its potential. He asserted that work to provide additional residential units should be started now before development costs become prohibitive. The city's study interviewed 420 firms, representing more than 6,500 jobs. The results of the study show that from 1969 to the early 1980s there was a 35% loss in jobs. In the past several years, however, many of the remaining firms have expanded with a resulting increase in jobs.

Used by firms

Only five area firms currently use the canal regularly, two oil firms and three concrete mixing plants. Other firms also use the canal, but less regularly.

The firms that continue to use the canal do so despite difficulties brought about by a buildup of silt along the bottom, decaying bulkheads and the wrecks of long abandoned barges.

By far the largest of the businesses on the canal is Bayside Fuel Oil Depot Corp., located at 510 Sackett St. According to the figures of company President Alfred Alligretti, Bayside distributes more than 200 million gallons of oil each year, making it necessary for barges to move through the canal every day.

Oil barged in

"We move all of our oil into the depot by barge, but we have several problems with the condition of the canal, particularly the miserable operation of the drawbridges and low water in the canal," said Alligretti.

He expressed hope that things will improve when the Douglass St. pumping station, which was closed due to lack of maintenance, is reactivated, creating a current. He expressed irritation that the pumping facility has been allowed to remain inactive for more than a quarter of a century.

Alligretti added that there are several hazards to navigation along the length of the canal, including the abandoned barges, and cars which have been dumped there.

5 bridges passed

He said his problems are compounded because his firm is only a little more than a block from the end of the canal, making it necessary for his barges to have to travel the entire length.

"We pass all five bridges spanning the canal, he noted, "Because we can only come in on the high tide in the morning, if there is a problem with one of them, we are out of luck," Alligretti maintained.

In two weeks, the beginning of a project to renew the canal will begin. Expected to be completed in 15 years, it will see a total rebuilding of the waterway.

Gregory Perrin, of the Department of Environmental Protection, said the clean-up will start with the refurbishing of the Gowanus pumping station. The job is slated for completion in 1988 and is expected to cost about $11.5 million.

Renewed station

The renewed pumping station will provide a current to keep water circulating through the canal and prevent the stagnation that spawns the odors. Farther down the line is a complete rebuilding of the canal, which will include dredging, new bulkheads and a new bottom for the waterway.

After completion, the canal will be fenced off to prevent the illegal dumping which contributes so heavily to its polluted state. With a completion date set so far in the future, Perrin said it was impossible to put a price tag on just how much the entire project will cost.

To the Editor, AMERICAN CANALS:

I thought the ACS Bulletin readers might be interested in the accompanying article as published in the Daily News for Jan. 19, 1986. The Gowanus is probably the only working Canal anywhere that can be reached by subway.

If you are not familiar with the Gowanus area, the canal's mouth is right next to the Erie Basin, the southern terminus of the Erie Canal. The five bridges mentioned are quite varied and include a double leaf Bascule, single leaf Bascule, and 2 swing bridges with off-center pivots so that the span rests in an opening in the canal wall parallel to the canal when it is opened.

The canal is also responsible for the very high viaduct immediately after leaving the Brooklyn-Battery tunnel, the viaduct had to be high enough to allow the Hamilton Avenue bridge (right under it) room to open. The Hamilton Ave. Bridge is a double leaf Bascule, but the roadway crosses the canal at an angle, so the spans are shaped like parallelograms instead of rectangles. And when the bridge is opened it looks quite strange with nothing square and all parts at different angles.

I'll keep you posted on further developments.

Charles F. Hruska
2074 Cropsey Ave.
Bklyn, NY, 11214

CHENANGO CANAL

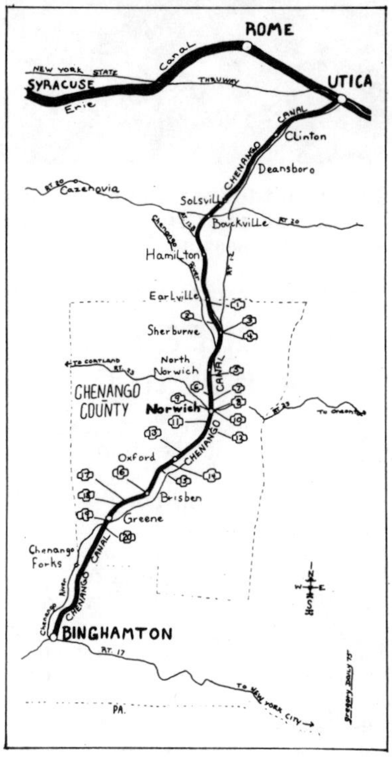

The Chenango Canal, officially opened May 6, 1837, ran 97 miles from Utica on the Erie Canal to Binghamton, on the Susquehanna River.

Pennsylvania eventually extended its North Branch Canal to Athens Pa., and cooperated in building the Junction Canal to Elmira, New York, which was operable by 1856.

The Chenango Canal then developed a connection with the Junction Canal at Athens, via the Susquehanna River, which became known as the Chenango Extension.

It was then possible for the boatmen on the Erie Canal to use this route to exchange goods (primarily anthracite coal) and passengers with Pennsylvania. The most notable voyage on the Chenango was a flotilla of 1,000 men of the 114th Regiment who left Norwich for the Civil War conflict, only half of whom every returned home.

The above map was provided by the Chenango County Bicentennial Commission. The numbers on the map indicate historical markers.

The Chenango Canal ceased operation in 1878.

HISTORY OF THE SHINNECOCK CANAL

Looking south toward Shinnecock Bay, where the water level is higher at this time of day. Hence, all gates are closed, except for the lift-lock gates at the left, where boats are entering and leaving the lock. Long Island Railroad Bridge in the distance.

QUERY ANSWERED

(Editor's Note: We had asked our readers some time ago if any of them could tell us anything about the Shinnecock Canal on Long Island. Responding to our query, **William L. Huber**, who lives at Sag Harbor, Long Island, made it his business to find out all about this tide-water canal, in considerable detail. He has done an excellent job of telling ACS members about the canal, its history, and its operation, in the accompanying articles, sketches and photos. If you have any questions after reading his material, and studying the sketches, we refer you to Bill Huber, himself, whose full address is: Box 333, RD 2, Whitney Road, Sag Harbor, New York 11963)

View from west bank of the Canal. Tidegates in the foreground. Lift-gates and control tower on the far side. (Huber photo).

By William L. Huber

In 1826 Holmes Hutchinson first proposed a canal to the state legislature for the purpose of accommodating boats and for reviving the dying marine life in Shinnecock Bay. He proposed locks at each end of the narrow isthmus which separates the inland waters of the South Shore of Long Island from the large bays which split the island into North and South Forks at Riverhead. He estimated the cost at $30,913.80. In 1828 the Long Island Canal Co. was organized with a capital of $200,000. Nothing came of this and in 1848 another company called the Long Island Canal and Navigation Company was organized with a capital of $300,000 with the same result.

Finally in 1879 the state legislature directed the state engineer, Horatio Swymour, Jr., to survey a canal to provide navigation and help in keeping the inlet open. Evidently there was an inlet of sorts at Shinnecock at that time but due to a lack of sufficient tidal flow it soon closed solidly not to reopen until nature accomplished what man could not on Sept. 21, 1938. He quickly approved the canal and estimated its cost at $35,000 with another $1,500 required annually to maintain it. By 1884 the legislature appropriated $12,000 as a start with two additional appropriations of $15,000 each in 1885-6. Later additional appropriations were made in the amounts of $22,000, $10,000, and $15,000 making a total of $89,000. The job was finally started and completed in 1892 but the canal was soon subject to intense scouring by the flow of water from Peconic Bay into Shinnecock Bay. The flow was in a southerly direction only being controlled by tide gates. Navigation could take place only during periods of slack water in the tidal cycle. The Long Island Rail Road bridge abutments were too close together and added to the problem until they were replaced by the present ones in 1896. In 1913 $46,000 was appropriated for construction of a lock to extend navigation time but a storm in 1914 ruined the work. In 1919 a new appropriation of $35,000 was made which, with $12,000 left over from the previous one, was given to the Town of Southampton to provide tide gates and a lock which was completed in 1919. Until the hurricane of 1939 reopened the inlet these gates were lashed open Firday to Monday for six months of the summer and during the entire winter. For 75% of the time water flowed both ways through the canal. In 1930 the state turned the canal over to Suffolk County with the Federal Government spending $420,000 during the 1930's for bulkheading and the construction of protective jetties on the Peconic Bay end. As soon as the inlet was opened in 1938 the gates were operated each day so that the flow of water was in one direction only from Peconic Bay into Shinnecock Bay and thence into the ocean through the new inlet. Since more water now flowed out through the inlet than in, it remained open although subject to shoaling.

On October 28, 1967 the present gates and lock were dedicated having cost $2.5 million with an additional $3.5 million being expended in recent years for bulkheading. Currently the House of Representatives has voted to release $200,000 to complete pre-construction engineering studies for stabilizing the Shinnecock Inlet, which with the Moriches Inlet to the west, is constantly shoaling up. The Army Corps of Engineers estimates that stabilizing the inlet will cost $18.1 million. Another $1.29 million has been set aside for maintenance dredging of the International Coastal waterway from Bellport Bay to the Shinnecock Canal.

In recent years the Shinnecock Inlet and the canal have grown in importance. The County recently spent well over a million dollars on a pier just inside the inlet to accommodate a growing fishing fleet. With the constant shoaling of the inlet due to the lack of stabilizing jetties, passing through the inlet is often extremely dangerous. Passage of boats through the canal has reached 28,000 per year with 7,000 passing in July alone. The necessity for keeping the inlet open and the increased traffic through the canal has made it one of the most important canals in the nation.

View from control tower, looking west. Tide gates visible across the canal (Huber photo).

HOW THE SHINNECOCK CANAL WORKS

By William L. Huber

Although sea level is constant, the distance from the inlets to the respective bays of Shinnecock and Great Peconic is considerable resulting in a 3′ variation in the water level depending on tidal and weather conditions. In order to provide a greater outward flow through the inlet than that provided by the incoming tide the gates of the canal are operated so that the flow of water is always to the south from Peconic Bay into Shinnecock Bay and thence through the inlet. Automatic tide gates 26′ in width close about 15 minutes after slack water due to the rising tide which takes 1½ hours to reach the lock from the inlet which is less than two miles distant. The lock gates are closed at slack water before the tide gates open in order to prevent slamming and damage. For the next 6 hours the waters of Shinnecock Bay will rise as much as 3′ over those of Peconic Bay before the next slack water period between the two bays. It is the function of the lock, which measures 40′ x 250′ with a height of 20′ from the sill to the top of the wall, to permit the passage of boats during this period.

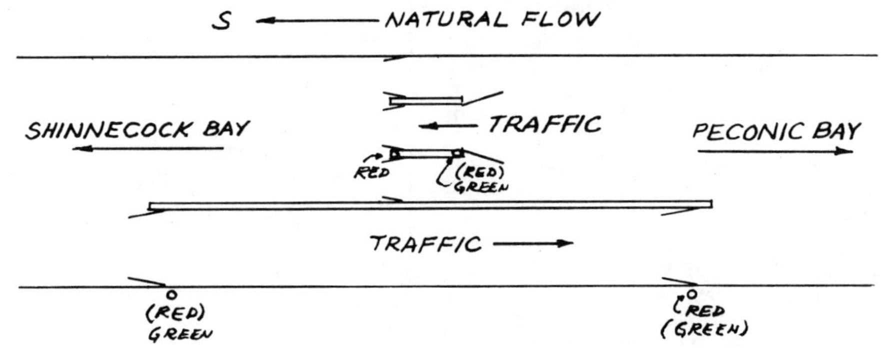

6 Hour Cycle When Peconic Bay Level is Higher Than Shinnecock Bay

When Peconic Bay rises above Shinnecock Bay the tide gates are pushed open automatically. The lock gates are kept closed briefly in order to force all of the water through the tide gates opening them more quickly. Once open the lock gates remain in the open position throughout this cycle. Southbound traffic is directed through the center tide gate by a green traffic light. The south light always remains red. Northbound traffic is directed through the opened lock bresting a current that can reach from 6-10 mph. A red traffic light on the north side of the lock protects the lock from potential two way traffic.

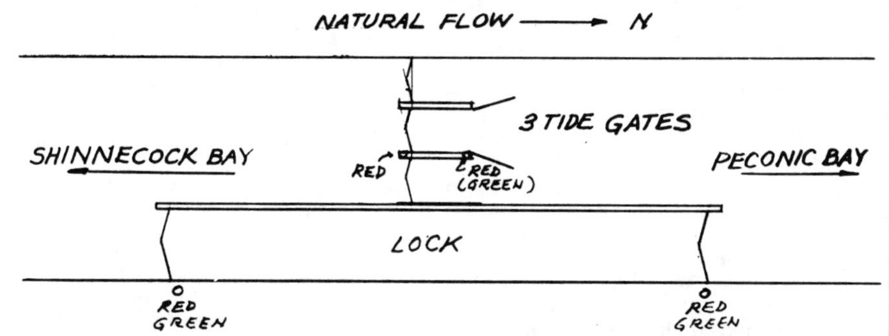

6 Hour Cycle When Shinnecock Bay Level Is Higher Than Peconic Bay

15 min. after slack water the rising waters of Shinnecock Bay close the 3 tide gates automatically. Prior to that, at slack water, the lock tender closes the lock gates. During this cycle marine traffic alternates through the lock directed by red and green traffic lights. The light on the south side of the center tide gate is always red there being no green light. The north light is also red during this cycle.

As soon as the tide in Shinnecock Bay starts to ebb through the inlet the water level recedes until eventually it falls below that of Peconic Bay, which is rising, causing a southerly flow which opens the tide gates. The lock gates are kept closed briefly in order to direct all of the flow through the tide gates hastening their complete opening when the lock gates are opened and left open for the remainder of this 6 hour cycle. It is this extra flow of Peconic Bay water which scours the inlet preventing it from closing again.

Traffic is controlled by a system of traffic lights mounted on each side of the center tide gate and on each end of the lock. The light on the south side of the center tide gate is always red, there being no green light. That on the north side is red when the gates are closed and green when the tide gates are open. Southbound traffic uses the center tide gate, which is fitted with funnel like bulkheads at the north entrance, and a red light on the north end of the lock directs traffic away from the lock. During the time all gates are open a green light on the south end of the lock directs all northbound traffic through the lock where it often must brest a 6-10 mph current.

As lock gates must have some means of dewatering them for repairs Shinnecock Canal has been provided with coffer dam leaves which drop into vertical slots on each side of each gateway, the leaves being stored adjacent to the lock. The lock gates are provided with slots on each side of each gateway so that a set of gates can be dewatered without dewatering the entire lock which is accomplished with portable pumps. During locking the flow of water in and out is by natural means through two of four 6′ dia. butterfly valves. Painted tide gauges on a board, one on each side of the tide gates, tell the lock tender the height of the water on each side. An additional gauge tells traffic of the clearance under the railroad bridge which has the lowest clearance. In addition the county provides facilities for lowering and raising the mast on sail boats, if necessary, at no cost. In the control tower a tide graph keeps a 5 week record of the tides. All operations are controlled from a console overlooking the tide gates.

It is an interesting operation to observe from the roads paralleling the canal on either side or from the Sunrise Highway bridge which crosses the canal just north of the lock. When the inlet becomes stabilized the canal will face ever increasing traffic from both the ocean and on the Inland Waterway. The canal is a vital link in the safe, inland passage extending from East Rockaway to Long Island Sound and Montauk Point.

SPECIAL EXHIBIT

Syracuse, New York: The Erie Canal Museum has mounted a new special exhibition which features first-person accounts that weave a rich tapestry of images of the Erie Canal, from its construction to its continued operation as part of the New York State Barge Canal System. The exhibition, appropriately titled, "Straight From the Horse's Mouth: Firsthand Accounts of the Erie Canal", is open June 15 - November 30 at the Museum, located in downtown Syracuse.

This portrayal of the Erie Canal through diaries, journals, ships' logs, newspaper accounts, oral histories, travel guides, and visual arts adds a new dimension to Erie Canal history.

Tolls may once again be charged on the New York State Barge Canal under a proposed constitutional amendment which would also permit the long-term leasing of canal-side property to private businesses. Whether the result will be the creation of a "Coney Island monster" and "environmental horror" or the revitalization of an underutilized, underfinanced, and seriously deteriorating state and national treasure, is a subject of controversy. The article by Connie Cohn appeared in a new newspaper, *Waterways:* New York's Waterfront News, March/April 1990.. The paper is published by North River Communications, P.O. Box 11, Croton-on-Hudson, New York 10520. A year's subscription costs $22.50.

THE JUNCTION CANAL

By Terry K. Woods

The Junction Canal of New York State was the final, vital link between the canal systems of the Empire and Keystone states, and as such, trundled thousands of tons of Anthracite, Plaster of Paris, salt and general merchandise between the two states from 1854 to 1871 - nearly 20 years.

As early as 1779, it is said, General John Sullivan suggested in a letter to George Washington that the southern waters (Chemung or Tioga branch of the Susquehanna river) be connected with those of the north (Senaca Lake). Washington is said to have presented the matter to the National Congress, but nothing came of it.

An attempt was made in 1815 to incorporate a private stock company - The Seneca and Susquehanna Lock Navigation Company - for $300,000 to build such a connection between the two waters. Large sums of money were hard to come by at that time in the private sector, however, (and for a long time to come) and the connection never materialized.

In the hectic and flush days following the instant success of New York's Erie Canal, most people not on the direct line of the Erie began clamoring for a branch of feeder canal of their own, and serious thought was once more given to a New York - Pennsylvania canal. In 1825, the same year the Erie was completed through to the Lake, James Geddes made a survey of the interstate connection and in his 1826 report proposed that a canal be built between Seneca Lake and New Town (Elmira) on the Chemung branch of the Susquehanna river.

Finally, after a great deal of haggling and opposition the 23 mile long, 48 lock Chemung canal was opened for navigation in October, 1833. In the meantime, the State of Pennsylvania had been literally inching a canal up along the North Branch of the Susquehanna toward the New York State line. By June of 1834 it was complete to the mouth of the Lackawanna - some 100 miles or so below Elmira.

In 1836, that part of the North Branch Extension from Athens south to Towanda was begun. Additional portions of the Extension were let in 1838. Thus, when the Extension would finally be opened for navigation there would have to be only eighteen or so more miles of canal constructed to finally effect that long awaited inter-state connection.

The residents of both states began petitioning their respective legislatures and clamoring for this short stretch to be started and rushed to a speedy completion. An engineering survey of the route was authorized in 1839 and detailed surveys of routes from Elmira to the state line were run along both banks of the Chemung river by Engineer Joseph D. Allen. His report was issued in 1840 and supported the feasibility of such a waterway along either side of the river.

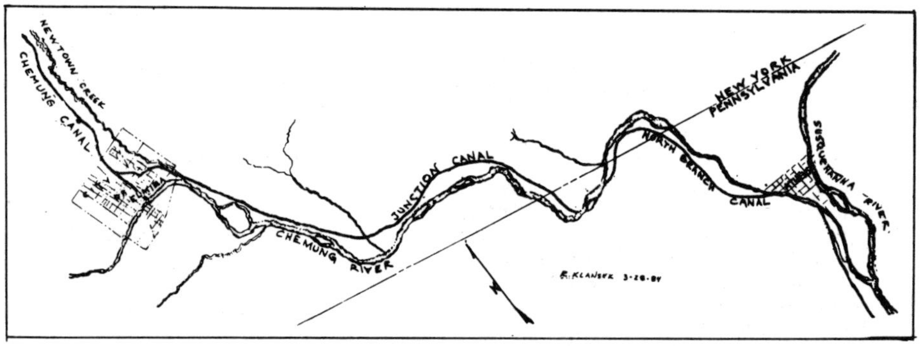

If economic conditions within the country had been normal, the Junction Canal would undoubtedly have been authorized by the New York State Legislature right then. However a serious economic 'panic' had struck the entire country in 1837 and hit again in 1839. In some portions of the nation cash money disappeared completely and all transactions were carried on by barter for a number of years. So devastating was this economic plunge that New York issued a STOP order on all new canal construction. Then, in May of 1841, the State of Pennsylvania halted all work on its unfinished Internal Improvements. The North Branch Extension Canal thus lay unfinished and useless. The head of navigation remained at Pittston on the Lackawanna.

For the next seven years nothing was done to connect the two state's canal systems. A private company was authorized by Pennsylvania in 1842 to complete the ill-fated Extension. In 1846 the State of New York authorized a private stock company - The Junction Canal Company - to incorporate for $500,000 and build the final canal link to Pennsylvania. However, the private sector in that part of the country was in no better position to finance a New York - Pennsylvania Canal than it had been in 1815, and both projects languished.

By 1848, economic conditions had improved to the point where the Pennsylvania Legislature again considered finishing the North Branch Extension and, in 1849, work on it was once more begun. An attempt was made to pay for construction of the Extension with receipts from the already operating canals. As a result, yearly appropriations were not nearly enough to keep a full force on the line during an entire season, and the work progressed very slowly. So great were the expectations that completed north-south canal system from the Anthracite fields of Pennsylvania to the frozen interior of New York would yield huge revenues, however, that in 1852, the State of Pennsylvania, against their previous resolutions, borrowed $200,000 and committed themselves to having the North Branch Extension Canal complete and carrying traffic by the 1853 boating season.

With this good news the Junction Canal Company had its New York charter renewed and obtained permission from the Pennsylvania General Assembly to construct their canal through parts of Pennsylvania as might be required to affect a direct communication with the Extension.

The actual construction of the Junction Canal began in March, 1853 at a point about three miles southeast of Elmira. Work was then initiated all along the line and by early 1854 most of the structures and channel were complete. It was June of that year, however, before boats actually navigated the Junction Canal between Elmira and Athens.

As originally constructed, the Junction Canal was 18 miles long with 7 miles in slackwater. It contained 11 wooden locks 17′ x 90′ in the chamber, 3 dams and 2 aqueducts. Minimum dimensions of the channel were initially 26 feet wide at the bottom, 40 feet wide at the water line and 4 feet deep. This allowed boats of 75 tons average capacity to use the canal. Around 1864, the minimum depth was increased to 4½ feet and the channel width to 42 feet at the water line.

One of the first boats through the new canal was the Packet, ST. LOUIS, Captained by Wheaton Loomis. His original intentions were to run the Packet between Elmira and Pittston, but the lower part of the North Branch Extension was in deplorable condition and it was impossible to maintain a head of water sufficent for continuous navigation for yet another two years. The Junction Canal also had a few water retention problems and Captain Loomis had to be content shuttling between Athens and Towanda on the North Branch Extension.

By late October, 1855, however, the Junction was again able to handle through traffic and Athens merchants began receiving goods by water from Elmira, Albany and New York City. Still, though, that section of the North Branch Extension between Tunkhannock and Pittston remained impassable. The Junction Canal, though open and carrying traffic, was not able to provide what it had been constructed for - access to Pennsylvania's Anthracite region and its canal system.

Official Pennsylvania State Reports claim that the two systems were finally connected in the fall of 1856 when two boats loaded with Pittston Anthracite negotiated the Extension and Junction Canal and arrived in Elmira on November, 17. That is essentially true, but the feat wasn't nearly as easy as it sounds. Before daylight on Tuesday morning, November 11, the "TOWANDA" under Captain A. Dennis and the "RAVINE COAL COMPANY NO. 4" under Captain T. Knapp - each loaded with 25 tons of Anthracite (each boat was capable of carrying 75 to 80 tons) left Pittston and carefully crept up the canal trying to find the channel. Since the channel only had a water depth of 25" careful steering was called for and the progress was understandable slow and tedious. These conditions persisted for the first ten miles or so. In addition, breaks in the canal banks caused several long delays and resulted in the tail boat becoming distantly seperated from the TOWANDA.

At McKune's Lock, some three miles above Buttermilk Falls, the depth of the channel bottomed at 3 feet. 'Sailing' became a lot smoother then and the TOWANDA reached the aqueduct just outside Tunkhannock early on Wednesday morning. There the crew of the maiden boat on the North Branch Extension was escorted into town by the village band and an impromtu celebration was held. The TOWANDA didn't stay long, however, and managed to reach the Horse Race by nightfall. They made the village of Laceyville early the next day, Thursday, where Captain Dennis was presented with the Stars and Stripes. He flew this proudly from his craft for the rest of the journey.

Browntown was reached later on Thursday and the Captain decided to wait there for the rest of his 'fleet'. It, the delinquent NO. 4, arrived just as daylight was making itself known the next morning. The channel here was broad and deep and it was decided to transfer the other boat's Anthracite into the TOWANDA. This long and tedious undertaking was finally accomplished and the TOWANDA 'set sail' to the north.

At Wyalusing they were greeted by a shouting throng and cannon fire. The cannon, itself, was presented to the TOWANDA's Captain and he used it frequently the remainder of his journey to announce their coming to the anxious popoulace along the banks of the canal. At Ulster the Packet ST. LOUIS passed them in the channel and received a salute from the Freighter's new cannon.

On the Sabbath, Captain Dennis had serious doubts about continuing on his journey during the Lord's Day. Upon reflection, though, he figured that completing the connection between New York and Pennsylvania's canal systems had become almost a religion in the area so it would be alright to continue, and he went on through Athens into the Junction Canal.

They were within one mile of Elmira and the completion of their long journey by Monday Morning. There they received word to wait until the townspeople could prepare a proper welcome. The Daily Gazette issued an EXTRA calling out the citizens to escort the first boat-load of Wyoming Valley black diamonds into town. The canal boat, and even the mules, were decorated and draped with bunting and flags. Then, at 2:00 in the afternoon of Monday, November 17, 1856, a procession formed before Haight's Hotel and, led by Wisner's Band, proceeded out to the boat, filling the deck to overflowing when they all finally arrived. A manned gun pulled by horses followed the crowd, shaking the air with its reports and receiving similar blasts from the TOWANDA'S own gun. The entire throng then entered town and the boat was halted at the intersection of the Junction and Chemung Canals where Colonel G. S. Hathaway addressed the crowd with an elegant speech welcoming the TOWANDA to New York's borders as the harbingers of another tie of brotherhood between the Keystone and Empire States.

The TOWANDA then entered the basin where the procession reformed and returned to Haight's where additional speech making and merry making went on and on.

Other boats followed the TOWANDA and, before winter closed the Junction Canal for 1856, 1,150 tons of Anthracite headed north out of Pittston toward Elmira. Leaks, breaks, and flood damage continued to plague that section of the Extension from the dam at the Horse Race down to the Lackwanna Aqueduct however and only 2,274 tons of coal were shipped in the entire 1857 season.

The State of Pennsylvania sold all its branch canals in May of 1858 and a private Company, the North Branch Canal Company, acquired the Extension and part of the old Wyoming Division from the New York State Line to Wilkes-Barre. The new Company repaired the feeder dam at the Horse Race and brought in tons of clay to repuddle the lower section of the Extension. They then leased the Junction Canal on a yearly basis and set about running a canal company. The whole line was open for navigation in July of 1858 and shippers managed to send 38,947 tons of Anthracite north before the end of that season, and 51,914 tons during the 1859 season.

The Junction Canal then settled down to a period of fairly active commerce, though annual flooding and the resultant damage to canal structures continued to plague the owners, operators and shippers. In 1863, at the height of the Civil War, the Junction Canal took in $25,764.93 while disbursements for maintainence, wages, etc. were only $9,419.23. Anthracite and Bituminous coal from Pennsylvania accounted for more than 80% of the total tonnage carried that year. Significantly, though, even in that rather flush year, no dividends were paid to investors.

The North Branch Canal suffered a devastating flood in 1865. The Company spent over $150,000 in repairs and it was October before it was again able to feed coal in to the Junction Canal. That same year, the Lehigh Valley R. R. gained a controlling interest in the North Branch Canal Company and obtained permission from the State Legislature to build a railroad along either the berm bank or the towpath as long as the waterway was maintained for navigation. The Company's name was changed to the Pennsylvania & New York Canal and R. R. Company and their line was opened all along the route of the canal by September, 1869. That spelled the end for both the Junction and North Branch Canals. Anthracite tonnages on the North Branch fell to 78,187 tons and on the Junction to 65,899 tons for 1869. Receipts on the Junction for that year were $13,253.35, or nearly $3,000 less than the expenditures.

Severe flooding on the Chemung in 1871 played havoc with the Junction Canal and its Pennsylvania terminus at Athens. Navigation wasn't opened until August of that year and only 511 tons of Anthracite made their way to Elmira. The Company's report for 1872 was short and sour. "This canal was not opened this year, and will not be." And so ended New York's Junction Canal.

THE CHEMUNG CANAL

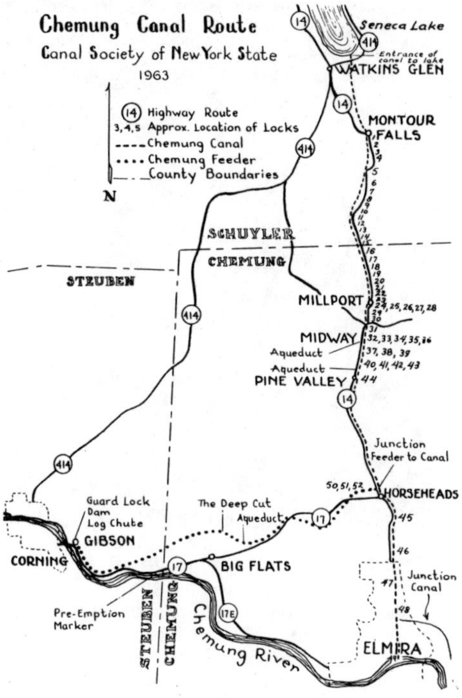

With the opening of the Cayuga and Seneca Canal in 1828, extending from Geneva on Seneca Lake to Montezuma on the Erie Canal, Erie Canal boatmen could travel as far south as Watkins Glen.

As Pennsylvania's North Branch Canal struggled northward plans developed for a canal to connect Watkins Glen with the Chemung River at Elmira, and later a Junction Canal to link Elmira with Athens, Pa.

The 25-mile Chemung Canal, built by New York State, was opened to Navigation in October 1833. It contained 48 locks, mainly to overcome a rise of 443 feet between Watkins and Pine Valley.

The Junction Canal finally made a connection with the Chemung Canal at Elmira in 1856, thus making it possible for anthracite coal to be shipped north to all parts of upper New York State.

However, severe flooding (and railroad competition) which plagued the North Branch, Chemung and Junction Canals finally led to the closing in 1872.

CAMILLUS—ERIE CANAL TOWN PARK

Volunteers from the town of Camillus, New York are shown moving some of the original stones of the Ninemile Creek Aqueduct on the Enlarged Erie Canal, 1836-1862. This now restored aqueduct (towpath shown here) was built in 1844, located between Locks 50 and 51, and was 144 feet in length. It was placed on the National Register of Historical Places in 1976. The accompanying map shows its location in connection with the restored canal and the town of Camillus.

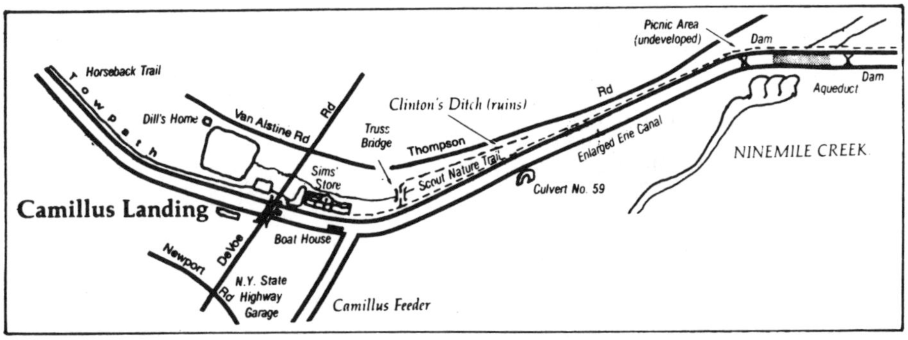

The Camillus-Erie Canal Town Park is a fine example of what a dedicated group of canal buffs (about 100 in number) can do in restoring and maintaining a historic canal section and turning it into an active tourist attraction.

Camillus is situated approximately five miles west of the outskirts of Syracuse, New York, along the route of the original Erie Canal, built in 1825, and also the Enlarged Erie Canal, begun in 1836 and completed in 1862, for a total length of 351 miles. Local volunteers, led by Dr. David Beebe, have been working for the past five years to make a rewatered section of the Enlarged Erie a truly working part of the local historical scene, as well as a picturesque asset to the community.

With financial support from the community and staffed by volunteers, the Camillus-Erie Canal Town Park offers canal boat rides on two separate boats along a two-mile rewatered section of the Second Erie Canal every Sunday from May through October. They have also rejuvenated the Sims Canal Store, which has been turned into a museum. Their two boats, the first named "Camillus-Erie" and the second called the "Ontario" offer rides along the Canal for a nominal fee. The "Ontario" is shown in Issue Number 50 of American Canals, and was launched in April of 1984 after considerable rebuilding and motor-mechanical work to get it running again. The boat itself has an interesting history.

For full information on the project, and a brochure, write: Town of Camillus, Erie Canal Project, 109 East Way, Camillus, New York 13031.

NEW MAP OF D.& H. RAILROADS

Delaware and Hudson Canal enthusiasts will be interested to know that a revised map of the two Gravity Railroads serving that canal is now available. Ed LeRoy, whose book *The Delaware and Hudson Canal* is now in its second printing, was the consultant on map revisions. It shows profiles of both the up-bound and down-bound tracks of Delaware and Hudson Canal Company Railroad, and the same for the Pennsylvania Coal Company Railroad. Both Gravity Road routes are shown in detail (with each plane identified) over the entire area of the northeastern Pennsylvania coal fields and canal junctions which they served. The 17" x 22" map, on good stock, is printed in three colors and is on sale at the Wayne County Historical Society, 810 Main Street, Honesdale, PA 18431.

ERIE LOCK 12 DEDICATION

By Gerard Chapman

Lock No. 12 of the Erie Canal was formally named and dedicated on May 19, 1990, as the "James Shanahan Lock," honoring an engineer and superintendent of the historic waterway. The lock is at Fort Hunter, N.Y., between Amsterdam and Fonda, where Mr. Shanahan lived.

James Shanahan (1829-96) was born in Ireland and brought to America at age 8. He became a stonecutter and assisted his brother, a contractor, in building Lock 50 of the canal of that era, as well as Locks 9 and 10 of the tributary Oswego Canal. He also worked on the famous Starucca Viaduct of the Erie Railroad, in Pennsylvania. Of it, Stewart H. Holbrook says in "The Story of American Railroads" (Crown Publishers, 1947), "the Starucca Viaduct, a 1200 foot stone affair" was "one of the great engineering feats" of its time which, "In 1947, a century after it was built . . . still carries the Erie's main line and looks good for another century or two."

As a partner of his brother and others, he built the masonry on parts of the New York Central and Oswego railroads and in 1854 built locks of the St. Mary's Falls Canal — "The Soo."

In 1864 Mr. Shanahan built the first bridge over the Hudson River at Albany, and from 1868 to -70 was superintendent of Section 3 of the Erie Canal.

The new historic marker at Lock 12 of the Erie Canal. (Photo by James C. Shanahan)

Elected to the Assembly (the lower house of the Legislature of New York) in 1869, he served on its canal committee, and in 1878 was appointed assistant superintendent of public works, having charge of the eastern division of the State canals.

Corollary to the dedication of the lock, was a dedication ceremony at Tribes Hill of Sacred Heart Church, which in 1875 had been enlarged and remodeled by Mr. Shanahan and his neighbors.

Several descendents of James Shanahan were present at the ceremonies; the brothers Thomas L. Shanahan of Calhoun, Georgia, and James C. Shanahan of Great Barrington, Mass., are grandsons; and the latter's granddaughter, Brynne Shanahan Kennedy, age 6, presented her great-great-grandfather's shillelagh to the New York Historical Society.

NEW BOAT FOR THE ERIE CANAL

Via **John Hulchansky** and **Bill McKelvey** we have secured the above photo of the "Nightingale II" from Michele A. Brown, its new owner, and president of Riverboat Cruises, Troy, New York. Mr. Brown's account of the boat's history and how he brought it home are as follows:

"Nightingale II" was built in 1954 at the Nolan Shipyard in Erie, Pa. Her dimensions are 64' x 17' x 4' and, as you can see, she is ideal for canals as her height is only 15' from the water at the stack. As far as I know, her original use was as a passenger ferry for the Straits of Mackinac and she has spent most of her time in the Great Lakes and canals. Passenger capacity is 90.

However, a year or so before I bought the vessel her Erie, Pa. owner moved to south Florida with the boat and then sold her to the Chattanooga Riverboat Co. in Tennessee. They used her for their 1984 season to establish a location in Chattanooga while planning and building a 400-passenger vessel. I bought her in October of 1984 and in February began the trip back to upstate New York. We travelled west on the Tennessee River, going downstream through the many Corps of Engineers locks, one of them 94' in a single locking. At Florence, Alabama we turned south and travelled the just-opened Tennessee-Tombigbee Waterway, the first commercial passenger vessel to pass through, and thence the Black Warrior River to Mobile, Alabama, the Gulf-Intracoastal Waterway to Carrabelle, Florida, across the Gulf of Mexico to Ft. Myers, Florida, across the Okeechobee Waterway to St. Lucie, up the Intracoastal Waterway to Norfolk, Virginia and on up to New York City and the Hudson River, north 160 miles to our home base on the eastern end of the Mohawk River/ Erie Canal. The trip was 3000 statute miles and took 40 running days.

From our base in Crescent we take passengers on daily sightseeing trips, passing through Lock 7 of the Erie Canal, a 27-foot lift. Evening cruises feature a prime rib dinner. I hope that we may meet all of you one day aboard the "Nightingale II."

Riverboat Cruises' address: 201 Second Street, Troy, New York 12180; Telephone (518) 273-8878.

SEMINAR

Roberts Wesleyan College in Rochester, New York, was host to 35 Elderhostel students for one week beginning July 27, 1986. The entire week was devoted to the study of the Erie Canal

The first course, "Literary Classics of the Canal Era" was the reading and discussion of works of Bryant, Hawthorne, Poe, and other authors of the era. These classes established a background of social and religious, as well as literary and intellectual attitudes of the period.

The second course, "Sing Those Canal Tunes", used for its text "The Canaller's Songbook", written by Dr. William Hullfish, and published by the American Canal and Transportation Center. Dr. Hullfish was a guest at one session. He sang for us, and recounted his adventures in tracing the words and music of these special folk songs. He also patiently and goodnaturedly listened to the canal songs written by the class.

The third section of study was "The Erie Canal: Clinton's Ditch Revisited". This presented the history, engineering, politics, problems and triumphs of the canal. A field trip for this course took us on a bus ride along the canal from Spencerport to Lockport, seeing the canal towns. At Lockport, we boarded "The Whalebird," locked through locks #34 and #35, viewed the old Lockport "five," sailed on to Tonawanda. The day ended with a picnic and sightseeing at Niagara Falls. *Elizabeth J. Gulino*

THE "OLD ERIE"

A photo, made around 1890, looking down the double-flight of locks on the old Erie Canal at Lockport, New York.

THE MOTT HAVEN CANAL

Old photo of the Mott Haven Canal, built about 1850 by Jordan Mott, inventor of the first coal-burning stove. (Courtesy "Yankees" magazine)

The following item was discovered by one of our ACS members in the June 28th, 1984 Issue of "Yankees" Magazine. It was written by the Bronx County Historical Society.

Almost everyone has heard of Canal Street in Manhattan, but how many Bronx residents know that there is a Canal Place in the Bronx, and that the story of the Bronx street is even more fascinating than the tale of the one in Manhattan?

The story begins in 1828. In that year, Jordan L. Mott, who had invented the coal burning stove (a major invention in the 19th century) opened a small iron foundry, (south of Yankee Stadium) at 134th Street and the Harlem River.

Over the years, the foundry grew in size, so that the original two buildings were dwarfed by the additions to the complex. By 1906, the business grew to be so large that there was not enough room on the original spot to expand, and the iron works moved to Trenton, N.J.

In any event, Mott saw possibilities in the expansion of the section in which his iron works had been built back in 1828, and as the first step in the development of the section, he purchased 200 acres at $175 an acre from the owner, Gouverneur Morris II, Mott immediately changed the name of the property from Morrisania to Mott Haven.

About 1850, Jordan L. Mott thought it a good business idea to dig a canal to replace a small stream that ran from 144th Street to the Harlem River between Third and Park Avenues. Mott dug his canal, but only to 138th Street, and it was called the Mott Haven Canal.

In 1864, Mott sold the land on either side of the canal to a man named Bryant. Four years later, Bryant began to extend the canal to 144th Street, with the understanding that there would be a bridge built over the canal at 138th Street to allow street traffic to cross.

However, Bryant sold his land to Rider and Conkling, who wanted to finish work on the canal. This aroused opposition from the residents of the Town of Morrisania, who felt the canal would be a nuisance and a source of malaria. Rider and Conkling pledged to maintain a bridge at 138th street, to dredge out the canal and bulkhead it, to build and keep in repair other bridges crossing the canal, and to fill in the canal at their own expense if it should become a nuisance.

The Mott Haven Canal was extensively used by the lumber yards and coal dealers which lined its banks. Shipments of fuel to the North River Electric Light and Power company came via the Harlem River through the canal to the company's power plant.

All this did not prevent the Town of Morrisania, and later the City of New York from declaring the canal a public nuisance, and after a full 25 years of litigation in the courts, the city began filling the waterway in 1901. A bulkhead was built at 138th Street in 1903, and the portion of the canal from there to 144th Street was opened as Canal Place, the Mott Haven Canal being allowed to exist below that point. The landfill was provided by the material excavated from 149th Street for the subway.

With the building of the Triborough Bridge in the 1930s, most of the rest of the canal disappeared. The lower portion of the Major Deegan Expressway was built as an approach to the new bridge, and this cut the canal at 134th Street.

Today, Canal Place runs from 134th Street to 144th Street in place of the Mott Haven Canal.

GENERAL LAFAYETTE REVISITS AMERICA, 1824

Submitted by William Dzombak

In the evening, we dined with the French residents of New York, who wished to celebrate the 47th anniversary of the battle of Brandywine. The dinner was prepared at Washington Hall [in New York City], and this patriotic and family festival was marked by a character both happy and original; many Americans, who were present, were struck with astonishment. Upon the very broad table a plan of the grand canal, in relief, was exhibited, which traversing the state of New York, unites Lake Erie with the Atlantic. This novel sort of chart occupied a length of sixty feet upon the table, where it was hollowed out of the thickness of the wood, and lined with lead; an extremely limpid water filled the canal, which was bordered with green sod, imitating meadows, in the midst of which were models of houses, trees, and animals. Bridges beautifully crossing the stream, and masses of rock under which the canal passed, and forests through which it wound along, all combined to complete this unique masterpiece of topography. Over the center of the table was a large sun in continual rotation.

Honorary Diploma

On the morning of the 18th of September at 8 A.M., Dewitt Clinton with a large number of citizens visited general Lafayette's apartments [in Albany], to present him in the name of the literary and philosophical society of New York with a diploma, constituting him a member of that society. On this occasion Mr. Clinton pronounced an elegant discourse which affected the general more, as coming from an orator who was the son and nephew of two distinguished men, with whom he had been intimately connected during the war of independence. During this short ceremony, a numerous procession was formed in front of our hotel, and at nine o'clock we embarked amidst the thundering of cannon, upon the canal which leads to Troy. Five handsome canal boats had been prepared for our voyage. The first carried a band of musicians; general Lafayette was in the second, with governor Yates, the ex-governors Lewis and Clinton, the mayor and city council, and some of the principal citizens of Albany. In the last three followed the escort commanded by major Coles, composed of an artillery company and three companies of infantry. . . . . .

At Troy, N.Y.

Before mid-day we arrived at a point where the canal communicates with the Hudson opposite Troy. At the sight of this town . . . general Lafayette was struck with astonishment. "What!" cried he, "this town has arisen as by enchantment!" "No", answered some one near him, "but it has been created and peopled in a few years by industry protected by liberty." The general then informed us that when in 1778, he crossed the Hudson at the same point, with a part of the army he commanded, there was then but two or three little cabins on the spot, in one of which, with some difficulty, he procured a cup of milk, and some corn bread. While he was giving us these interesting details, our boat descended to the river, where a dozen decorated boats towed us across to Troy.

Levasseur, A.
Lafayette in America, vol. 1, p. 95, 117
Philadelphia: Carey and Lea (1829)
Research Reprints, NY (1970)

AMERICANS AND CANADIANS HOLD JOINT MEETING

Perfect weather was provided for us the week-end of May 15-17 when the American Canal Society and Canadian Canal Society met in Syracuse, New York, for a combined tour and meeting.

It was the Fifteenth Anniversary of the founding of the American Canal Society and the Fifth Anniversary for the Canadian Canal Society.

The Wine and Cheese Party on Friday evening was attended by most of the registrants; the rest joined us for the Saturday activities. The Friday program included remarks by both Bill Trout, President of the American Canal Society and Bill MacLean, President of the Canadian Canal Society. Bill Shank concluded the evening with a slide show of canal and river trips he had made in USA, Canada and Europe. Canal memorabilia and books were on display.

One of the groups assembles for a photo at the entrance to the Erie Canal Museum in Syracuse. Guide – Debbie Stack.

We inspect "Mud-Lock", a well-preserved relic of the Old Oswego Canal, on the north outskirts of Syracuse. Keith Kroon, ACS Tour Chairman, left center of the bridge.

One of the CCS-ACS groups inspects a canal boat replica at the Erie Canal Museum in Syracuse. Guide – Don Wilson.

A model of the weighlock mechanism, where whole canal boats and their cargos were "put on the scales".

After breakfast at the Ramada Saturday morning, the group boarded two buses and made their first stop at the Canal Museum in downtown Syracuse. Here Debbie Stack, Public Relations Director for the Museum and Don Wilson, Museum Curator, took charge of the group and guided them through the various exhibits at the Museum, including the new full-scale Erie Canal Boat, built into the Weigh-Lock. ACS Vice President Bill Gerber presented the Museum staff with a multipage volume of detailed canal-index sheets for the USA and Canada which had been worked up over the past ten years by the ACS Canal Indexing Committee. The buses then moved on to "Mud Lock" on the old Oswego Canal, and next to Lock Number One on the currently operating Oswego Canal at Phoenix, New York. Our lunch stop was made at the Lock Three Restaurant in Fulton, where we watched a number of boats "locking through" while enjoying our buffet lunch. We visited several additional locks on the Canal, before arriving at the H.L. White Maritime Museum in Oswego, New York, located on a jetty extending out into Lake Ontario. Rosemary Nesbitt, the Museum Director, was our guide.

(Continued on Page 44)

AMERICAN – CANADIAN JOINT MEETING

Lock Number One on the presently-operating Oswego Canal presents an array of draw-bridges to the water-traveler.

The Erie Canal Weighlock Building, showing part of our group touring full-scale canal boat, built into the old weighlock chamber.

(Continued from Page 43)

Back at the Ramada in North Syracuse, the Canadian Canal Society had a brief business meeting, while the ACS members enjoyed a "Happy Hour". Our evening banquet was served on time, and we enjoyed a full hour of canal music by Dr. Bill Hullfish and his "Golden Eagles String Band" — also a fine talk on the "Erie Canal, Past and Present" by Don Wilson of the Erie Canal Museum. A definite "plus" for the evening was a slide talk on restoration work being done on the British Canals by Jonathan M. Morris, an English Lockeeper who was visiting the USA courtesy of the Winston Churchhill Memorial Trust.

Sunday morning, after an excellent buffet breakfast at the Ramada, the ACS-CCS groups again boarded their buses for a visit to the Rome, New York area. Enroute, we were treated to a running lecture by Howard Card, a local canal buff, as we traveled for nearly thirty miles along the "Old Erie Canal".

Arriving in the Erie Canal Village, we first rode the "Independence" mule-drawn canal boat, which took us along a short stretch of the "Old Erie" and then a narrow-gauge steam train, interspersed with visits to a fine canal museum, as well

A small boat leaves a Lock on the Oswego Canal, bound upstream. Note the dam and power house.

Robert Legget, author of several books on the Canadian Canals (left) visits with Louis Cahill, American Canal Society Director for Canada. Lock Number One on the Oswego Canal in the background.

Interior of the H. L. White Maritime Museum at Oswego. Lecturing to our group (left center), Rosemary Nesbitt, Museum Director.

Our group rides the "Independence", a mule-drawn canal boat replica on a restored section of the "Old Erie". The mule-team may be seen on the towpath at the far right.

as a carriage museum, livery stable, blacksmith shop, weaving craft house, old church and school house, amongst other historic displays.

After a light lunch at the Village "snack bar" we headed for Fort Stanwix, in downtown Rome, built by the British in 1758. This Fort was rebuilt by the Americans in 1776 and withstood a 21-day seige by the British in 1977. The Fort has now been restored to its original condition and is being operated by the National Park Service.

On the top deck of the "Independence" at Erie Canal Village, as we ply our leisurely way along the "Old Erie".

At Silvan Beach, we rode Captain Gradel's small "Canal Cruise" boat along part of the present Erie Canal and out into Lake Oneida.

A narrow-gauge steam train took the ACS-CSS group on a four-mile ride along the Rome and Ft. Bull Railroad at Erie Canal Village.

The Canadian group left us at Fort Stanwix for an early start back to St. Catharines, Ontario. The remaining group headed for Silvan Beach on Oneida Lake where we rode Captain John Gradel's canal boat, along a portion of the modern Erie Canal and out into the Lake, amidst a huge crowd of Sunday tourists and boaters enjoying the first nice week-end in May.

In charge of the bus tours was Keith Kroon of Rochester, New York. Responsible for the "contacts" with all the Canadians involved was John Burtniak of Thorold, Ontario. We had extremely good cooperation from the staff of the Ramada, the bus company, the Canal Museum and all the other agencies along the way — a most successful week-end!

Dr. Bill Hullfish (Author of "Canallers Songbook") entertained us Saturday evening, with his "Golden Eagles String Band". The group played and sang a number of old canal ballads for us.

THE OSWEGO CANAL

This old etching shows the Oswego Canal joining the old Erie Canal in downtown Syracuse. On the left is the raised towpath crossing over the Oswego entrance. The Erie Canal (now Erie Boulevard) disappears in the distance under the "Lock and Mulberry Street Bridge." The old Weighlock Building (now a museum) is at the right.

by Debbie J. Stack

The Erie Canal might have extended from Albany to Oswego rather than Albany to Buffalo, had politicians from Oswego had enough voting power to control the waterway's route. Although western New York's growing communities eventually won the battle to extend the Erie west from Syracuse to Buffalo, the Oswego interests refused to give up, petitioning the state legislature throughout the construction of the Erie to connect Oswego with the soon-to-be famous canal.

Their persistence was rewarded. In 1819 the Canal Commissioners recommended that a one mile side-cut be constructed from the Erie Canal in Syracuse north to the salt-making operations at Salina. This move in Oswego's direction was followed by authorization, in 1822, to extend the side-cut to Onondaga Lake and eventually to dig a water connection between Onondaga Lake and the Seneca River.

Things looked very promising when, in 1823, the Oswego Canal Company was chartered to build a canal to Oswego. Progress was slow, however, and in 1825 New York State appropriated $160,000 for the Oswego Canal and took over the company's work.

On July 4, 1826, the 50th anniversary of the signing of the Declaration of Independence, a ground-breaking ceremony was held in Fulton, New York to celebrate the initiation of construction of the final section of the Oswego Canal, the 12 miles from Oswego Falls to the Oswego Harbor. The 18 pound cannon used to mark the occasion exploded prematurely, but caused no injuries.

A fever epidemic among diggers delayed the work, but the Oswego Canal was completed in December, 1828, and on April 28, 1829 opened to navigation throughout its 38 mile expanse. The 4 foot deep and 40 foot wide Oswego Canal was constructed at a cost of $565,437.35, more than twice the original estimated cost of $227,568.33.

According to Noble Whiteford in his **History of the Canal System in the State of New York,** 19.7 miles of the Oswego Canal were independent canals connected with the Oswego River by locks and dams, with the remaining 18,56 miles consisting of slack water river navigation accommodated with a towpath along its bank. A total of 22 bridges, 14 locks, 1 aqueduct, 8 dams, 7 culverts, and 2 waste weirs were constructed on the canal, providing for a total descent to Lake Ontario of 155 feet, from Syracuse.

Competition between Oswego and Buffalo rose anew. With the completion of the Oswego Canal boats could, at Syracuse, either take the Oswego Canal to the port of Oswego and transfer cargo to lake boats for travel on Lake Ontario and the Welland Canal to Buffalo Harbor, or continue on the Erie Canal to Buffalo. The Erie route was more costly. The Oswego Canal offered savings in both tolls and time. There were no tolls levied for travel on Lake Ontario and the time delays encountered at the Erie's 22 locks between Syracuse and Buffalo were significantly cut on the Oswego Canal. Boats taking the Oswego route saved approximately two days over the Erie route.

By 1850 the Oswego Canal was more profitable per mile than the Erie and in 1862 it was enlarged to 7 foot deep and 70 foot wide. The traffic jam of boats waiting to be weighed at the Syracuse Weighlock, located at the junction of the Erie and Oswego canals, was so great that it was decided that two Erie-bound boats should be weighed for every one Oswego-bound vessel. This did not significantly reduce the jam, however, so in 1866 an Oswego Weighlock was built.

Boats travelling west from Oswego were laden with salt, manufactured goods, lime, gypsum, iron, timber, and coal. Those travelling east from Oswego carried cargos of agricultural products, including grain, flour, cheese, butter, lard, wool, and pork.

The Oswego Canal, like the Erie, was adversely affected by railroad competition. In addition, after the 1883 abolishment of canal tolls, the Oswego Canal's financial competitive edge over the Erie Canal was also reduced, thus cutting traffic even further on the Oswego Canal.

With the opening of the state's Barge Canal System in 1918, the route of the Erie Canal moved northward, utilizing Oneida Lake and other natural waterways. The Oswego portion of the Barge Canal System began at Three Rivers Point, the junction of the Oneida, Seneca, and Oswego Rivers (rather than at Erie Boulevard and Oswego Street in downtown Syracuse.) It travelled north (approximately 25 miles) following the Oswego River, and terminated at Oswego Harbor. There are presently seven locks, overcoming a total drop of 118.5 feet. The Oswego Canal became second only to the Erie Canal in terms of usage.

Debbie Stack is Public Relations Coordinator for the Erie Canal Museum.

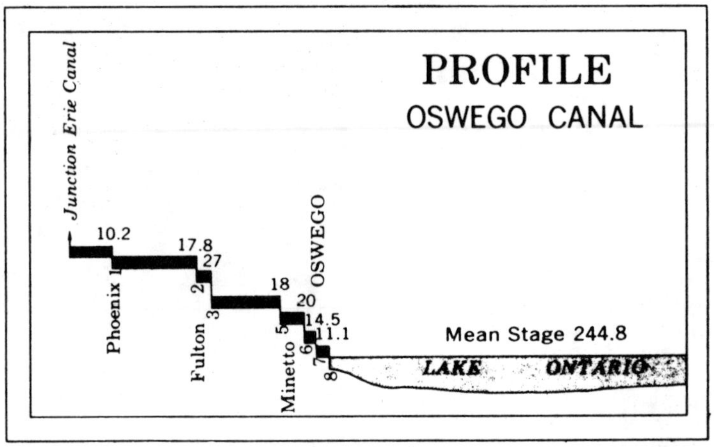

1792 BATTEAU RECREATED

The "Discovery" — replica of an Eighteenth Century Batteau, such as used on the Mohawk River and early waterways west of Albany. Overall length: 30 feet, 4 inches. Maximum width at the gunwales: 7 feet, 3 inches. Carrying capacity: 1.5 tons.

ALBANY, N.Y. — New York State Museum technicians are building the first replica ever constructed of an 18th century Mohawk River batteau as part of a two-year statewide celebration of the birth of the Canal Era in New York. The batteau, named DISCOVERY, will be on exhibit at the Urban Cultural Park Orientation Center at the Schenectady Museum throughout the summer, and will participate in reenactments and living history programs planned at towns along the modern Barge Canal.

"Batteau" is a French term for "boat" which came to signify, 200 years ago, any flat-bottomed, shallow-draft vessel that was pointed at both ends. This vessel was the mainstay of inland shipping, particularly for the military, until the end of the 18th century.

Batteaux (the plural) came in different sizes, known generally as 3-handed, 4-handed or 5-handed according to the crew needed to propel them. There were undoubtedly many variations in design, but all were characterized by a flat bottom made up of pine boards laid lengthwise, with battens nailed across to hold the bottom together. Oak frames, usually made from natural crooks, fastened the bottom to the pine planks that formed the sides of the vessel.

These craft were propelled by poles and oars, with a small sail used when the wind permitted. The Mohawk River batteaux built in Schenectady were apparently smaller and lighter than most, because of the shallow and often constricted channel they had to navigate and the several portages around which they had to be carried on their way west.

For the past several years the State Museum has been researching the contributions made by the Western Inland Lock Navigation Company (1792-1820) to the opening of westward transportation in New York. This private company was chartered by the New York Legislature in 1792 to find ways to improve the inland water route from Schenectady to the Great Lakes. Among the improvements made were the creation of several short canals, with dams and locks, that bypassed obstructions to navigation, thus issuing in an age of canal travel decades before the better known Erie Canal crossed the State from Albany to Buffalo.

The core of the Museum's research—known as "The Durham Project"—has been the discovery of archaeological remnants of these historic engineering works and the development of a plan for their preservation. Publications will result from this research when it is completed. In the meantime, this educational program is designed to bring some of the research findings to public attention during the bicentennial years of the building of the works which were the true beginnings of the Canal Age in New York.

In order to enhance public interpretation of this era of inland transport, a typical commercial batteau of the type built in the 1790s in Schenectady for the Mohawk/Oneida route has been recreated. Using archeological remains of earlier military batteaux in State Museum collections, historical accounts and drawings, and general technical data on small boats of the period, a probable design was determined. Construction duplicates original materials and techniques where possible.

This vessel is meant to represent the new, 3-handed batteau purchased on August 20th of 1792 by the Western Inland Lock Navigation Company from a boatyard in Schenectady for the survey of the Mohawk River, which was the first field operation of the newly formed company. The replica will serve as a "vehicle" for several bicentennial programs staged on the modern Canal, including a reenactment in Schenectady of that 1792 Mohawk expedition. It will also be incorporated into exhibits and programs at various regional museums during the bicentennial.

ERIE CANAL BOARD GAME

On winter evenings, when the canal is frozen solid, both young and old game board players will enjoy learning how to captain a freight boat from Albany to Buffalo and back, and make a profit, in spite of the many hazards, regulations, and accidents that a boat captain must know how to deal with.

The game board is a map of the Erie Canal, printed on stiff paper, that is 5x34 inches. The map needs to be glued or taped to a rigid backing (cardboard). The pieces or tokens that move on the game board are two boats (folded paper cutouts), one boat westbound (from Albany) and one eastbound (from Buffalo). Movement of boats is determined by the spin of an arrow mounted on a numbered cardboard circle. Each player starts the trip with a stack of paper money used to pay fines (as for speeding, or running a boat that is overweight). Profit made on a trip is calculated by subtracting expenses paid out (tolls, fines) from income realized by selling (auctioning) the cargo at the end of the trip. The cargo carried on each boat is determined, at the start of the game, by drawing "cargo cards" from a stack of face-down cards; the auction price of each item of cargo is found by spinning the arrow pointer.

The game is sold by Chatham Hill Games, P.O. Box 253, Chatham, NY, 12037. The price is six dollars, which is a lot to pay for a large sheet of paper and a plastic spinner, but students of canals are known to attach unusual value to the subject that interests us all.

THE BLACK RIVER CANAL

By Bruce J. Russell - Contributing Editor

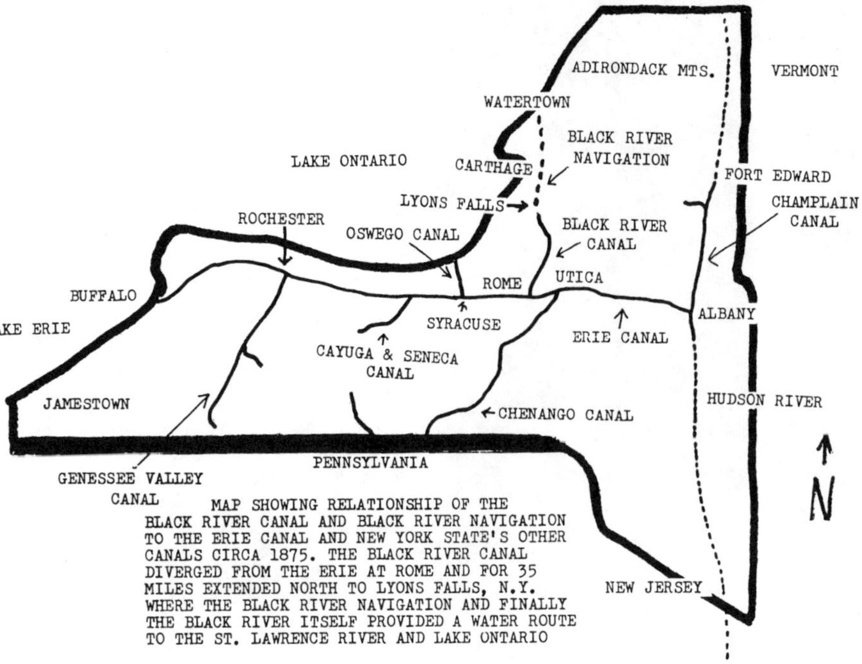

MAP SHOWING RELATIONSHIP OF THE BLACK RIVER CANAL AND BLACK RIVER NAVIGATION TO THE ERIE CANAL AND NEW YORK STATE'S OTHER CANALS CIRCA 1875. THE BLACK RIVER CANAL DIVERGED FROM THE ERIE AT ROME AND FOR 35 MILES EXTENDED NORTH TO LYONS FALLS, N.Y. WHERE THE BLACK RIVER NAVIGATION AND FINALLY THE BLACK RIVER ITSELF PROVIDED A WATER ROUTE TO THE ST. LAWRENCE RIVER AND LAKE ONTARIO

On Oct. 8, 1994 the CANAL SOCIETY of NEW YORK ran a field trip exploring the remains of the BLACK RIVER CANAL which closed in 1926. A small segment of it continues to function as a conduit for bringing water to the NY STATE BARGE CANAL. Water from the Black River was collected behind a dam, and sent via a feeder canal to the summit level of the Black River Canal at Boonville, N.Y. It then was used to 1) Operate the Black River Canal from Boonville to Rome, and 2) To supply the Erie Canal's summit level, also in Rome. While some water reached this summit directly from the Black River Canal, most of it took another route. South of Boonville a feeder tapped into it and brought some water into a natural stream called the Lansingkill. The Lansingkill took this water first to the Mohawk River which then fed into the Erie Canal near Rome. While a lot of water was diverted from the Black River Canal enough was left to operate its locks. In 1918 the NY STATE BARGE CANAL replaced the ERIE. However water from the Black River was still brought via almost the same route to the successor waterway. The Black River originates in the hills above Rome and flows north into the St. Lawrence River near Watertown, N.Y. It was the main source of water for the Erie Canal from Rome to Albany.

On the weekend of October 8th and 9th the CANAL SOCIETY of NEW YORK STATE, hosted a two day field trip to the remains of the BLACK RIVER CANAL, one of the Empire State's lesser known but nevertheless interesting waterways. The BLACK RIVER CANAL was selected by the trip committee because it features 109 locks in the relatively short distance of 35 miles, making it a "climbing canal" par excellance. Unlike the longer ERIE CANAL which comprised numerous level segments completely devoid of locks, the BLACK RIVER CANAL began in Rome where it connected with the ERIE and proceeded north into the hills and low mountains which in the 19th and early 20th Centuries were sources of lumber. In many of the rural villages which the waterway traversed the stately homes of these "lumber barons" are still visible and bear testimony to the economic muscle which gave impetus to the digging of the canal.

The building of the BLACK RIVER CANAL for navigational purposes was preceded a few years earlier by construction of a non-navigable feeder from the Black River to the ERIE CANAL in the vicinity of Rome, N.Y. The purpose of this channel was to transport water from that mountain stream to the summit level of the ERIE CANAL to keep it filled.

It was later decided to construct a navigable canal extending from Rome northward into the Black River country where it would terminate in Lyons Falls, a distance of 35 miles. From Lyons Falls to Carthage the actual Black River itself would be improved and used for navigation with some essential structures added such as locks around rapids and slack water dams. North of Carthage the Black River was navigable without the need for any artificial improvements. The actual BLACK RIVER CANAL therefore extended only from Rome to Lyon's Falls. From Lyons Falls to Carthage, also approximately 35 miles, the waterway was known as the BLACK RIVER IMPROVEMENT. Surveying for the BLACK RIVER CANAL began in 1834 and was finished in 1837 when contracts were awarded for the construction of the initial 14 miles from Rome into the "North Country" which was rich in timber and other resources. 20 years later, in 1855, the job was finished.

Once it was opened the BLACK RIVER CANAL was used to transport lumber, both in the form of logs as well as boards produced at the numerous saw mills of such canalside communities as Boonville and Lyons Falls, to the growing cities of New York State such as Syracuse, Buffalo, Albany, and of course New York City. The canal boats took several days to work their way south through the 109 locks to Rome where they continued east or west on the "mainline" ERIE CANAL. Because of the time involved in passing through the locks there wasn't any passenger service. The construction of railroads north of Rome in the direction of Watertown and the Canadian border in the 1850s, 60s, and 70s spelled the end of large traffic volumes on the BLACK RIVER CANAL. Plans to extend it northward to either Lake Ontario or the St. Lawrence River were shelved by the time the waterway was completed in 1855. Nevertheless during the late 19th Century it remained active and old photographs show boats loaded down with timber descending its multiple flights of locks.

OCTOBER 7 - EARLY BIRD SPECIAL TOUR

The CSNYS field trip was scheduled as a weekend affair, however, for the benefit of those able to take Friday Oct. 7 off there was an "early bird" or preliminary series of activities connected with the BLACK RIVER CANAL. It was a "drive yourself" automobile tour of selected locations which would not be covered during the main Saturday Oct. 8 bus trip. Those taking this option were given a guide who directed them from Rome north to Carthage where navigation on the BLACK RIVER CANAL and BLACK RIVER IMPROVEMENT had its northern terminus. (From Carthage tugboats were able to travel over the unimproved

We stop at Lock #16 at Westernville. This was a well-built lock, still in good condition. (Russell photo)

Black River to Watertown and the St. Lawrence River).

From Carthage the participants traveled south in the direction of Rome. Various sites of interest were identified such as sawmills and tannerys. One of the most impressive was the remains of Basselin's sawmill which when built in 1885 was the largest in the United States.

At Forestport there was a stop at the dam and start of the 11 mile feeder canal which takes water from the Black River directly to the summit level of the BLACK RIVER CANAL at Boonville. A dam was constructed across the Black River at Forestport, thereby creating a pool of water. From this pool an 11 mile mini-canal (which was navigable) brought water to Boonville, the highest point on the BLACK RIVER CANAL. This same water then traveled south on the BLACK RIVER CANAL to Rome where it supplied the ERIE CANAL'S summit level.

The data sheet and directions for the Oct. 7 "Early Bird" Special Tour were prepared by CSNYS member Howard Card.

On the evening of Oct. 7 an excellent slide show was presented by Craig Williams at the BEECHES, headquarters for the weekend extravaganza. The show concentrated on the history of the BLACK RIVER CANAL and the region it served.

Delta Dam, just north of Rome, which provided a large reservoir for both the Black River Canal and the Erie Barge Canal. The Black River Canal by-passed the dam on the triple locks at the right and then crossed Delta Creek in the foreground. (Old photo courtesy of Paul & Kate Bartczak)

OCTOBER 8th BUS TOUR

Saturday October 8th dawned bright and sunny, ideal weather for a journey over the BLACK RIVER CANAL and through the North Country. A total of 85 people signed up, two bus loads. The tour leaders were Howard Card and Craig Williams.

Eight stops were scheduled with a lunch break in the middle. This was what could reasonably be covered in a day excursion and which would give participants a good idea of what the BLACK RIVER CANAL was all about.

Upon departure from the Beeches our two motor coaches proceeded north into the Black River country. Passing Delta Lake and Delta Dam constructed in 1910-1918 to supply water to the new BARGE CANAL we saw the abandoned prism of the BLACK RIVER CANAL. The BLACK RIVER CANAL brought water directly into DELTA Lake until it was abandoned in 1926. However since the BARGE CANAL still required a supply of water an elaborate feeder system was relied upon. Water continued to flow through the abandoned canal from the summit level at Boonville to a point a few miles south of that community. Water was then diverted into a natural stream called the Lansingkill which took it first to the Mohawk River and then into the BARGE CANAL in the vicinity of Rome. This complete rerouting of the canal's water supply permitted total abandonment south of Boonville.

Lock #16 is an example of excellent masonry on the Black River Canal. (Russell photo)

We stopped at Lock #16 in Westernville. Here was found some of the earliest style of lock construction on the BLACK RIVER CANAL. This consists of poor quality stone rather than the neatly formed perfectly square blocks used on other waterways and other portions of the BLACK RIVER CANAL. From Boonville, the summit level, all the way to Rome the original construction of the lock chambers was sub-standard. This may have been done in an effort to save money. Anyhow repairs and rebuildings were necessary throughout the life of the canal. Lock #16 exhibits portions made from the original stone plus sections which were rebuilt about 1906 using concrete. This lock like most of those on the BLACK RIVER CANAL had a lift of 10 feet and dimensions equal to those of the original ERIE CANAL, sometimes called "Clinton's Ditch". The locks were never made wider as they were on the ENLARGED ERIE CANAL which replaced the original 1825 waterway in the 1840s. Adjacent to Lock #16 is a building which once served as a canal side store. A structure situated several hundred feet away was once thought to be the lock tender's house, albeit enlarged and remodeled following relocation from the actual lock. However some people doubt that it served in this capacity. Who knows?

Departing from Lock 16 we once more proceeded north along the highway, passing additional locks, many covered with dense foliage but still visible. The tour guide pointed out locations of a number of aqueducts which carried the BLACK RIVER CANAL across streams. All are now missing their wooden water troughs and only the stone abutments remain. Driving through the hamlet of North Western, N.Y. the guide identified the location of a canal boat yard and shop. Lock 19 was viewed without stopping but the guide pointed out that it had been completely rebuilt a couple of years before the canal was abandoned in 1926. Neatly cut stones were used to replace those of the earlier 1830s construction. Historians now know that some of these well formed stones were actually taken from locks of the abandoned CHENANGO CANAL elsewhere in NY State and transported to the BLACK RIVER CANAL for use on repair and reconstruction work. It therefore appears that at least until the 1920s there was a plan to keep this waterway in operation.

Howard Card, one of the CSNYS Tour Guides, demonstrates operation of the water control gates to the Black River Canal prism. (Russell photo)

STOP #2 THE FIVE COMBINES

This is one of the most spectacular sites on the BLACK RIVER CANAL - the so called flight of five locks arranged on a "staircase." Such positioning of locks allowed the canal to quickly gain elevation. On the BLACK RIVER CANAL many staircases exist. Less than a quarter of a mile from the "Five Combines" is a smaller one consisting of three chambers in sequence. The FIVE COMBINES are Lock numbers 39 to 43 and the nearby flight of three locks are numbers 44-46. Both of these typify this unique waterway. Our buses parked adjacent to the site and we walked a short distance before encountering them. Our guidebook contained photographs showing these locks in operation during the early part of the present century. Lock houses were likewise visible. An examination of the individual chambers indicates that several rebuildings and reconstructions occurred during the life of the waterway. Stones from the abandoned CHENANGO CANAL were utilized in the 1880s. In 1912 yet another rebuilding

(Continued on Pages 50-51)

THE BLACK RIVER CANAL

Our lunch stop was at the HULBERT HOUSE at Boonville still serving fine food and drink - 68 years after the canal is gone. (Russell photo)

(Continued from Page 49)

occurred, this time using concrete as well as better quality blocks. Nothing survives of the tenders' houses once situated adjacent to these abandoned lock chambers which are filled with trees and other vegetation. It was difficult for me to believe that as late as 1920 barges and canal boats were still passing through here. The site of the two staircases has been proposed for restoration by Oneida County. A canal era park is envisioned. For this to occur considerable funding will be required to stabilize the lock chambers. On the BLACK RIVER CANAL all locks had wooden flooring which results in less stability than those utilizing stone.

Control gate atop old Lock 70, which was rebuilt using concrete in the final years of navigation on the canal - 1926. (Russell photo)

STOP #3 - THE FEEDER SYSTEM

We then visited the junction of the BLACK RIVER CANAL and the Lansingkill Stream or small river which carries water to the existing BARGE CANAL near Rome. Originally water was taken from the Black River at Forestport and sent via an 11 mile navigable feeder canal to Boonville where it supplied the summit level of the BLACK RIVER CANAL. It then traveled for 5 miles through the canal's prism to a point above Lock #70. Here is was diverted into an artificial non-navigable channel called the Lansingkill Feeder which led directly into the actual Lansingkill which in turn brought it to the Mohawk River and finally to the ERIE CANAL's summit level at Rome. After 1918 when the BARGE CANAL replaced the 19th Century ERIE the same arrangement continued. One of the reasons for this was that had ALL water been allowed to travel through the lower section of the BLACK RIVER CANAL which joined the ERIE CANAL at Rome a great deal would be lost to seepage since the soil in that location was very porous. Hence the needed water was removed at a point below Boonville and sent via a natural stream to Rome. Of course enough water was allowed to remain in the BLACK RIVER CANAL for it to continue operation. However when the BLACK RIVER CANAL was abandoned in 1926 it was decided to permit all the water to flow through it for several miles and then to empty into the Lansingkill at a point where it runs parallel to the canal prism. The result of this is that for a few miles the prism of the BLACK RIVER CANAL retains water. The Lansingkill Feeder is still intact and the control gate leading to it is located just above Lock #70. If need be water can still be sent through it.

All of these channels and feeders illustrate the point that canaling involves much more than simply digging ditches and filling them with water. Canal building was a science developed over hundreds of years - first in Europe and later in America. Always an adequate supply of water for a canal's summit level must be secured, often from points not situated on the actual navigation channel.

On top of Lock #70 is a metal house containing a water control device which regulates the amount of water flowing from one portion of the BLACK RIVER CANAL to another. After passing through the control apparatus this water flows for a few miles through the canal and then dumps into the Lansingkill Stream which brings it to the Mohawk River and then the BARGE CANAL. Our two guides HOWARD CARD and CRAIG WILLIAMS explained to the tour participants just how this feeder system worked.

STOP #5 - LUNCH AT THE HULBERT HOUSE

Lunch was served at the historic HULBERT HOUSE, originally built in 1812 using local stone. This building therefore predates by 43 years the completion of the BLACK RIVER CANAL. The meal which was a buffet featuring "all you could eat" would certainly qualify as "canalers' fare." There was chicken, meatballs, spaghetti, salad, and several kinds of cakes, pies, and puddings for dessert. After walking many miles along dusty towpaths and opening and closing countless locks canalers normally developed ravenous appetites. Establishments such as the HULBERT HOUSE made certain these men never left the premises hungry.

STOPS 6-8 and CONCLUSION

Following lunch the buses headed north along Route 12 in the direction of Lyons Falls where the 35 mile man made channel terminates. Several locks were passed, some with only one wall remaining. It then became a divided highway with a wide grass covered medium separating the northbound and southbound lanes. From this point it was obvious that the road occupied the space of the abandoned BLACK RIVER CANAL prism. Lock 85 was encountered which now sits in the center of this median strip.

STOP #6

At this location a flight of four locks is situated in the center of the aforementioned Route 12 highway median. These four chambers date from 1850 and are Lock numbers 87, 88, 89, and 90. In 1968 when the highway was rebuilt and widened to its present configuration it was necessary to build directly atop the abandoned canal prism. Rather than destroy the lock staircase it was decided to let it remain within the median as a monument to N.Y.'s Canal age. Protective fencing has since been erected for safety reasons and historic markers posted telling the story of the BLACK RIVER CANAL. Our tour stopped here for 15 minutes. CRAIG WILLIAMS explained that this "canal park" was one solution to the problem or preserving old lock chambers. However he cautioned that on a long term basis it did nothing to

The dam across the Black River at Forestville provided a reservoir lake for both the Black River Canal (through an 11-mile feeder entering the canal at Boonville) and the Erie Canal below. (Russell photo)

stabilize these structures and that eventually pressure would cause their walls to collapse inward.

STOP #7

We entered the town of Lyons Falls and drove next to the Lyons Falls Pulp and Paper Company. The parking lot of the industrial complex contains the remains of Locks 103, 104, and 105. Obviously this factory was built after the BLACK RIVER CANAL was shut down in 1926. Beyond this location that abandoned canal prism gradually descends to the level of the Black River. Locks 106, 107, 108, and 109 - the final one of the canal - are still intact and can be seen. Interestingly #106 and 107 were rebuilt in 1918, just 8 years before the BLACK RIVER CANAL was abandoned. Concrete was used in place of the original 1852 stonework. Once again the thought occurred to me that almost until its closure people were still convinced the BLACK RIVER CANAL was viable and were willing to spend money on it, rebuilding deteriorated lock chambers.

Lock #109 feeds directly into the Black River. According to our guides this is where tug boats attached themselves to the barges and towed them further north to Carthage and beyond to Watertown and the St. Lawrence River. The section from Lyons Falls to Carthage was known as the BLACK RIVER NAVIGATION because it wasn't an artificial channel with a towpath. Instead it consisted of the actual Black River with some improvements which permitted navigation. The BLACK RIVER NAVIGATION contains a few locks plus artificial channels around waterfalls but it's still a natural waterway and not a true canal.

STOP #8

The final stop of the tour was at Forestport where a 200 foot long dam stretches across the Black River, creating a lake. This dam was originally built in 1848 and rebuilt in 1902 and again in 1970. Its purpose was to collect water which would be sent south to Rome to supply the summit level of the ERIE CANAL. An 11 mile long feeder canal took water from the lake to Boonville where it entered the BLACK RIVER CANAL which allowed it to reach Rome. However some of this water was removed from the BLACK RIVER CANAL below Boonville and forwarded to the ERIE CANAL via the Lansingkill Stream. The aforementioned 11 mile Forestport to Boonville Canal was navigable, but the Lansingkill Stream wasn't since it was much too shallow. Although the BLACK RIVER CANAL was abandoned in 1926 water destined for the BARGE CANAL, successor to the ERIE CANAL, still runs from the Black River at Forestport to Rome via a short section of man made canal plus the Lansingkill Stream. However until this final closure the BARGE CANAL could also receive its supply of water directly from the lower section of the BLACK RIVER CANAL via natural flow through its locks. As stated previously this route was viewed as less efficient due to extensive loss through seepage along the length of the canal's channel. Therefore the Lansingkill was utilized.

Final Lock (#109) on the Black River Canal, at Lyons Falls. From this point north, navigation was in the Black River itself. (Russell photo)

We visit the control gate at Lock 70, providing water for the Erie Barge Canal. (Russell photo)

Locks 87, 88, 89 & 90 are preserved in the medial strip of Highway #12 just north of Boonville. (Russell photo)

Upon departure from Forestport the buses headed back to the Beeches where an evening banquet was planned. During the course of 8 hours the tour participants acquired an amazing knowledge of the BLACK RIVER CANAL. Its most interesting sites were visited and an explanation of its source of water was given. It was obvious that this waterway was one of the most spectacular not only in New York State but in the nation as well. One hundred and nine locks in the distance of 35 miles represents an amazing feet of engineering. For a few decades until the coming of the railroads and highways it served its purpose of transporting lumber from the "North Country" to the more built up portions of the state. Much of its cargo continued beyond Rome to points on the ERIE CANAL or even on some of the lesser known "laterals".

Following the banquet there was a presentation by DR. LIONEL D. WILD about Walter D. Edmonds and his book *ROME HAUL* which deals with life on the BLACK RIVER CANAL during its heyday. It certainly made the day's activities truly meaningful. Canals were more than water filled ditches and stone lock chambers. They were also people - men and women earning a living in an era still considered "frontier" and lacking many of the comforts we now take for granted.

The CANAL SOCIETY of NEW YORK STATE is to be commended for running an excellent excursion which required considerable advance preparation. The souvenir guide book was a genuine help in understanding the BLACK RIVER CANAL and its place in history. I look forward to future field trips with this group.

BIBLIOGRAPHY

CANAL DAYS IN AMERICA - The History and Romance of Old Towpaths and Waterways by Harry Sinclair Drago - Bramhall House, New York, 1972

ROME HAUL By Walter D. Edmonds - Syracuse University Press, 1987 (Originally Published 1929)

Our Tour Guides - Craig Williams (left) and Howard Card. (Russell photo)

New York State's Working Lock Model

The New York State Canal Corporation shows off its Operating Lock Model at a public convention. (Photo by David B. Cox.)

Canal people are always trying to explain how a lock works, so every canal museum and canal society needs a working lock model (i.e. with water) for demonstration purposes, even if there is a real one out the front door. This gives children and their parents a chance to be shown how a lock works, and ideally, to try it themselves without the danger of a real-life disaster.

At the International Conference on Historic Canals we learned of a successful portable working model used by the New York State Canal Corporation, which now runs the 524-mile state canal system under the NYS Thruway Authority. David B. Cox, Assistant Director, Office of Canals, sent us this photo and a description of the model, which can be set up at public events to show how a New York canal lock works.

The lock was made in 1987 by Mort Hollenbeck, an employee of the Canal Corporation in their sectional headquarters in Lyons, New York.

The lock chamber, made of aluminum, is 26″ long, 8-3/4″ wide with a depth of 11-1/4″; its lift is about 5″. The whole box is 82″ long, 20-1/2″ wide, and 14-1/4″ high.

NEW CHENANGO CANAL BOOK TELLS STORY OF THE 1837 WATERWAY

Hidden among the trees of the Chenango, Oriskany and Sauquoit valleys in central New York are remnants of the Chenango Canal which connected Utica on the Erie Canal and Binghamton from 1837 to 1878. The remains are monuments to the engineering genius of another century. The canal was designed by John B. Jervis, resident engineer on the Erie Canal. There are still lock walls, aqueduct ruins, culverts and hand-dug ditches to be found which have stood the test of time.

A new book, *Limestone Locks and Overgrowth: The Rise and Descent of the Chenango Canal* by Michele McFee (published by Purple Mountain Press of Fleischmanns, NY) directs readers to remains of the canal which are accessible near towns along its path (mostly routes 12 and 12B today), and reveals the history of this remarkable landmark.

The Chenango Canal was intended to reach the coal fields of Pennsylvania. As early as 1838, a survey was done along the Susquehanna River from Binghamton to Waverly for an extension which would tie the canal to a Pennsylvania system. Construction began in 1863 and stopped in 1872 when only 20 miles of the 40-mile extension had been completed. The book's chapter on this episode is the most comprehensive account of the subject to date.

The book, a large-format, quality paperback with 240 pages and 200 illustrations, is available for $25.00 at book and gift shops in Central New York or from the publisher, Purple Mountain Press, PO Box E3, Fleischmanns, NY 12430, 1-800-325-2665. Add $3 shipping and sales tax.

BOOK REVIEW

ALWAYS KNOW YOUR PAL: CHILDREN ON THE ERIE CANAL. *Edited by Debbie J. Stack and Donald A. Wilson of the Erie Canal Museum, Syracuse, New York.*

Reviewed by Bill Shank

A 103-page, 6″ x 9″ paperback with a 4-color photomontage cover, the book is a charming review of all activities involving children on the 19th-Century Erie Canal system.

The book contains more than fifty old photos, etchings, quotations, and drawings from such diverse publications as *St. Nicholas Magazine, Harpers Weekly, Marco Paul Adventures, McGuffey's Eclectic Primer,* and *Buffalo Bill's Wild West portfolio.*

This interesting publication successfully weaves together all the daily scenes and experiences of children traveling with their families on old Erie Canal Boats—the schools they attended (in winter), the books they read on board, the food they ate, the friends they made, the chores they performed—in short, just about all the daily experiences they enjoyed as Canal Boat Children.

To obtain a copy write: Erie Canal Museum, The Weighlock Building, 318 Erie Boulevard East, Syracuse, New York 13202. Price $14.85, plus $1.50 for shipping.

ERIE CANAL MUSEUM AT LOCKPORT

Lockport's Erie Canal Museum, at the bottom of the flight, was an abandoned power plant about five years ago when Lockmaster Joel Beyer decided to do something with it. He and his partner Lock Operator Ed Voltz scraped the walls, painted the building, re-did the windows, put up the sign and put canal things inside. Then others brought more memorabilia and displays for the museum and a carpenter built cabinets for it all. The latest addition, parked outside, is a small Buoy Boat which once refueled buoy lamps along the canal. In this photo, the Buoy Boat is in the center, flanked by the museum (right) and the old flight of five locks on the left, still spanned by picturesque arched footbridges made of stone slabs. When you go there, look for the rope burns in the iron railing, made by tow lines. It would be fun to have gates in the old flight again! (Bill Trout)

ADIRONDACK WATERS: LIFEBLOOD OF CANALS

by David L. Kipp

New York is one of the few states to have still in operation a direct descendent of an inland waterway built during the canal-building craze that flourished in the second quarter of the 19th century. At that time, canals (called "internal improvements" along with roads and railroads) were seen as ways to link together regions of our then-expanding country, making it easier and cheaper for commerce to flow westward. For example, Philadelphia wanted to be connected with Pittsburgh on the Ohio River; the state of Ohio, to connect its namesake river with Lake Erie as did Indiana by way of the Wabash River; Illinois desired to link Lake Michigan with the Mississippi by way of the Illinois River; New Jersey, the Delaware River with New York Harbor; and Maryland, Chesapeake Bay with the Ohio River. These and other states caught the fever and built canals or canalized rivers to bring produce or raw materials from the hinterlands to manufacturing cities and ports of export in hopes of generating an economic boom.

New York's Erie Canal, loinking the Atlantic Ocean via the Hudson River to the Great Lakes at Buffalo, was the only one that in one continuous pathway truly pierced the Appalachian Plateau between the eastern seaboard and the interior midwest. It was the first completed (1825) among the major long-distance canals and was an instant and long-term success. Its tolls not only paid for its own construction, operation, and upkeep, but also subsidized construction and maintenance of a half dozen economically unsuccessful branch canals in New York and an Erie Canal enlargement to keep up with growing traffic, which did not peak until after the Civil War. By then, the competition from railroads was sending all but a few of America's canals into bankruptcy and abandonment. The railroads first lured away passenger traffic, then light freight, and were making inroads into heavy bulk cargos. Initial improvements made on the Erie Canal attempting to prevent defection of shippers to railroads were: doubling of locks and widening of aqueducts to prevent bottlenecks; and widening and deepening of the canal bed to allow bigger barges to be used. Later, locks were lengthened to permit two barges to be locked in tandem; mechanical changes to the locks and gates were tried to reduce the time of each lockage. However, as the 20th century approached, New York, now desperate to keep freight on the canal, eliminated tolls and began a second enlargement. But it was soon realized that this was only a Band-Aid and that what was really needed was an end-to-end modernization to complete the transition from the age of human and animal power to that of steam and electricity. Thus the New York Barge Canal system was born—constructed between 1903 and 1918. It remains little changed today.

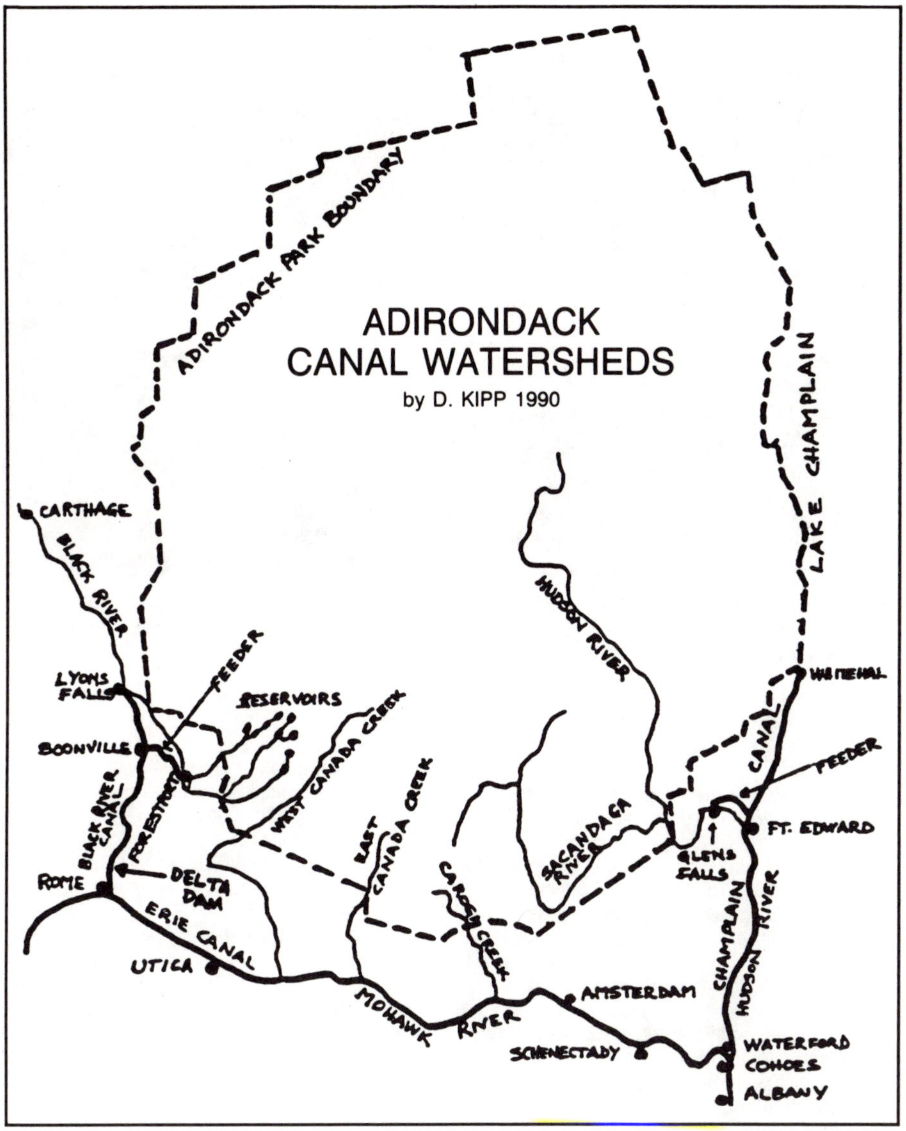

Black River Canal, Lock 64—top of the Boonville Gorge looking south. The canal here began a steep descent from the summit level down to the Mohawk River valley carrying water tapped from the Adirondack foothills ultimately to the Erie Canal at Rome.

Whether it was the original Erie (often called Clinton's Ditch for Governor DeWitt Clinton, its chief promoter), or the enlarged Erie which existed from the 1840s to 1918, or today's Barge Canal system, an adequate, dependable water supply was and is absolutely essential to canal operations. This is where the Adirondack watershed comes in.

A canal without an adequate water supply is like an automobile without gasoline—useless for its intended purpose. Erie Canal surveyors and engineers in their determination of the canal route considered not only the lay of the land but also water sources and feeders along the way, especially for the highest or summit levels.

In New York State, builders of the Erie and some of its branch canals were blessed by geography and topography with two major water sources that made the canal system possible: Lake Erie and the Adirondack Mountains. Lake Erie, which is over 500 feet higher than the Hudson River at Albany, supplies the western end of the canal with what would seem to be an infinite supply if it were not for the fact that it's not downhill all the way to Albany. After reaching a low point near Syracuse, it is necessary to surmount the divide between the Great Lakes and Hudson/Mohawk river watersheds in the vicinity of Rome, New York, before descending through the Mohawk valley to the Hudson River at Cohoes. It is to this summit level and the remaining section of the Erie Canal eastward that the Adirondacks supply their precious waters so vital to the Erie's success. The Adirondack streams supplying the Erie directly are East and West Canada creeks and Caroga Creek. In the past, the Black River and some of its tributaries also supplied water by way of the Black River Canal.

(Continued on Page 54)

ADIRONDACK WATERS: LIFEBLOOD OF CANALS

Delta Dam on the Mohawk River—a few miles north of Rome, this dam impounds the drainage from the western Adirondack foothills for today's Barge Canal. It replaced the old water delivery system by way of the now-abandoned Black River Canal.

(Continued from Page 53)

Two New York branch canals were also inextricably linked to Adirondack water for their operations—the Champlain and the Black River canals. Both received 100% of their water from the Adirondacks. The Champlain Canal lies in that ancient transportation corridor which links the Hudson River with Lake Champlain. The vital importance of this link to the north was highlighted by the French and Indian War and Revolutionary War battles that swirled along its route at Ticonderoga, Lake George, and Saratoga. It was recognized as a natural water transportation route in an early but unsuccessful attempt at a canal by the Northern Inland Lock and Navigation Company in the 1790s. So when the Canal Commission headed by DeWitt Clinton planned New York's original canal system, the Champlain was included as a branch canal to be constructed concurrently The Champlain was completed shortly before the Erie (1823) and its direct descendent continues today as part of the modern Barge Canal system. The Adirondack water it used came primarily from the Hudson where it was tapped above Glens Falls and directed eastward to the Champlain Canal in a 12-mile-long navigable feeder. Today, the Champlain uses the Hudson River itself as the canal from Albany to Ft. Edward and depends on the Glens Falls Feeder to supply its summit and the descent to Whitehall on Lake Champlain, thus continuing its dependence on the water released from the Adirondack Mountains.

Black River Canal

The Black River Canal is a bit of a different story. When the Erie Canal's early success brought boomtown status to such places as Buffalo, Rochester, Syracuse, Utica, and Schenectady, it generated considerable envy on the part of nearby districts where it was thought that a canal would also bring automatic economic prosperity. The people of the Black River valley, which lies on a north-south axis between the Adirondack foothills north of Rome and Watertown near Lake Ontario, were no exception. They petitioned Albany for many years for a canal link to the Erie and many surveys were taken and alternate routes considered. It was finally approved by the legislature in 1836 and construction began. The deciding factor for this authorization was not necessarily the persistence of the Black River settlers but the fact that the Erie Canal needed the water that a Black River Canal could deliver. Before the Black River Canal was completed, the Erie Canal would occasionally have to shut down in late summer and early fall when there was insufficient water to operate the Rome summit level. The Adirondacks, acting like a giant sponge, retain winter snow melt and spring rains, slowly releasing them at a steady rate throughout the summer and fall, supplementing the Erie's supply when linked by the Black River Canal.

The Rome summit level must have sufficient water supply to provide for lockages in both east and west directions (each lockful of water is eventually lost to the canal when it reaches the lowest point) and make up for losses. Even on the old Clinton's Ditch, each lockage required 125,000 gallons of water (for a typical lock of 15-foot width, 90-foot length, and a lift of 10 feet) and traffic often exceeded 100 boats per day passing a lock. This ongoing operational need plus additional losses due to surface evaporation, seepage, gate leakage, and overflow (some water must be allowed to flow over the waste weirs to maintain the canal at the proper level) meant that large volumes of water must continually be supplied to the summit level. Adirondack water from the Mohawk River supplemented by the Black River Canal was thus a key to maintaining canal operations.

Black River Canal Combine Lock—one of several sets of combines on this "mountain goat" canal which had 108 locks compressed into a mere 35 miles as it surmounted the divide between the Mohawk and Black River valleys. The sign reads, "Black River Canal—1837-1922—Connected Mohawk Valley with Black River country opening up 90 miles of navigable waterway."

The Black River Canal was begun in 1838, suffered through a stop-work period when New York State experienced a shortage of funds, and was finally completed in 1855. Its junction with the Erie Canal was in Rome. From there it paralleled the Mohawk River, branching off to follow the Lansing Kill up the Boonville Gorge across a tongue of the Tughill Plateau, arriving at its summit level at Boonville. It then descended into the Black River valley to meet the river at Lyons Falls. The river was by 1861 made navigable as far north as Carthage. The Black River Canal is justifiably called the "mountain goat" canal as it had 108 locks in 35 miles (3 locks per mile). Compare this to the original Erie's 83 locks in 363 miles (4.4 miles per lock). Many of the abandoned locks can still be seen along Route 46 between Rome and Boonville and along Route 12 north of Boonville.

Summit Level Supply

To supply the summit level of the Black River Canal, the Black River was dammed at Forestport and a 14-mile-long navigable feeder was dug from there to Boonville. This feeder can still be traversed today by canoe, and its towpath is an excellent hiking or cross-country skiing trail. Water in excess of the needs of the Black River Canal was allowed to flow into the Lansing Kill and from there to the Erie Canal by way of the Mohawk River. As traffic on the Erie increased during the middle of the 19th century, the Black River proved insufficient in dry years and several Adirondack lakes were dammed to make reservoirs that could release water when the Erie was in need. These bodies of water included Woodhull, Bisby, Twin, North and South Branch, Sand, Canachagala, and White lakes. This water supply route continued to function well into the 20th century when the Mohawk River was impounded by Delta Dam and Lake. These were built north of Rome to supply the much larger needs of the modern Barge Canal completed in 1918. By then the Black River Canal had already lost the battle with the railroads for shipping the region's lumber and ceased operations in 1925. It was because it was a channel for bringing Adirondack water to the Erie that it lasted as long as it did. Other New York branch canals like the Genessee Valley, Chemung, Crooked Lake, Chenango, and Onieda Lake canals were abandoned in the 1870s due to expenses far exceeding revenues.

It is interesting that Verplank Colvin, the renowned surveyor and cartographer of the Adirondacks, by virtue of his intimate knowledge of the headwaters of all the streams that exit the mountain massif, had strong ideas about the channeling of the mountains' vast water reserve. In his report to the legislature for 1874–79, he reiterates the canal system's dependency on the Adirondacks:

"In New York the great canals . . . are all of them, on their upper levels, dependent upon the Adirondack region for water supply; so that, but for the feeders from the streams of the wilderness, the canals themselves would be impracticable." [Verplank Colvin, Topographical Survey of the Adirondack Region: Report to the Legislature 1874–79 (Weed, Parson & Co., 1880), p. 329.] In an earlier report, Colvin promoted a remarkable visionary scheme for sending water from the high Adirondack lakes to the canal or waterway most in need (including the St.. Lawrence):

"The amount of water available from the wilderness for each of these canals (Erie, Black River, Champlain) is nearly equal, and may, by means of skillful engineering, be made extremely equitable in its distribution, in a manner which the peculiar character of the region renders possible, and which—though at first apparently astounding—is actually simple. The barometrical leveling of my survey in showing the altitudes of lakes has developed most important facts. It has been found that the interlocking headwaters of the rivers, flow often, for great distances in opposite directions and at inconsiderable differences of elevation; while the topography of the intervening country after examination of the defiles and ravines has been found to frequently admit of the construction of connecting canals; so that with aid of carefully arranged dams and reservoirs the water could be made to flow in either direction. Thus a single mind at the Capitol of the state, may control by telegraph the flow of numerous and considerable rivers; and at the tap of the key, turn so many million cubic feet of water from the St. Lawrence to the Mohawk, the Hudson, or to Lake Champlain! Such a proper system of reservoirs in this commanding position would prevent the stranding of boats in one canal for lack of that water which at the same instant might be swelling in floods through the affluents of the rivers feeding the others—by excess of rains on the distant watersheds—and would afford to each the quantity to which its commerce and its importance entitle it." [Topographical Survey of the Adirondack Region: Report to the Legislature 1873/74 (Weed, Parson & Co., 1874), pp. 104–05.]

Certainly, Colvin's reports emphasizing the rôle of Adirondack forests in supplying a steady flow of water to three of New York's canals were a factor when the 1885 law establishing the Adirondack Forest Preserve cited this as one of the reasons for maintaining the forests intact. Shippers who were in favor of the law may not have been so much devoted to the forests as interested in a healthy canal system that gave them a hedge against high railroad rates.

Black River Feeder Canal—Through this channel, the waters of the Black River and an array of mountain reservoirs provided enough water to operate the Black River Canal and keep the Erie Canal's summit level watered in dry years. It carried the water from a dam at Forestport 14 miles to the summit at Boonville. It is still navigable today, at least for canoes, and the towpath is in great shape for hiking and cycling.

The fact that Colvin's schemes of interconnected, telegraph-controlled headwaters never came to fruition does not diminish the ongoing rôle of the Adirondack watershed as the supplier of the lifeblood for New York canals, past and present.

[David L. Kipp is an engineer with Eastman Kodak. He is a member of the A.C.S. and several regional canal societies. His writing has appeared previously in American Canals.*]*

Black River Feeder Dam—Along with several upstream tributaries dammed as reservoirs, the water flowing off the shoulders of the southwest Adirondacks into the Black River drainage was here sent by a 14-mile-long navigable feeder to the summit level at Boonville. From there it descended through the Black River Canal joining the Erie Canal at Rome.

Canal Way Trail Albany to Buffalo

Excerpted from "THE CANALER," *Fall 1993 published by the Canal Corp., a subsidiary of the New York State Thruway Authority.*

A "Canal Way Trail" will stretch the length of the New York State Canal System, enhancing recreational opportunities for all canal system uses.

Plans for the trail were announced recently by Gov. Mario M. Cuomo and Thruway Authority/Canal Corporation Chairman Peter Tufo.

"The Canal Way Trail offers an exciting opportunity for recreational users to explore the beauty and history of the canal, which played such a vital role in the development of New York State and the Midwestern United States," Gov. Cuomo said.

The initial phase of the Canal Way Trail will extend and improve more than 120 miles of trails by 1995, increasing public access and opening up the canal corridor for land-based recreational use. This initial phase is on state-owned right-of-way. It may be a combination of paved trail and fine gravel, referred to as "stone dust."

The Governor noted that the trail system will connect cities and villages with agricultural areas and pristine open spaces, and also provide enhanced access to the cultural and recreational resources of the canal, including parks, museums and historic sites.

"Thanks are due to Canal Corporation Chairman Peter Tufo for initiating this program," Gov. Cuomo said.

Through cooperative efforts among state and local governments and volunteer trail organizations, five trail segments have been identified for initial completion. This first phase will include the development of new trails as well as enhancements in areas where trails are in deteriorated condition.

Work planned for this year includes about five miles of trail along the Old Erie Canal near Rome, where work on the canal began on July 4, 1817. This five-mile section will be constructed by Thruway maintenance forces.

Phase one improvements to existing trails will include: 37 miles of trail along the Old Erie Canal between Rome and Syracuse (portion between Rome and Canastota being designed this year, with work due to be completed by early next year); segments east and west of Rochester totaling 25 miles (including eight-mile segment between Pittsford and Perinton being designed this year, completed early next), and 16 miles between Lockport and Medina.

New trails to be constructed in the first phase are: more than 40 miles between Herkimer and Fort Hunter in the Mohawk Valley, and nine miles along the Glens Falls feeder canal, between Fort Edward and Glens Falls in the Champlain Canal section.

Trail planning will be coordinated closely with the on-going efforts of the State Office of Parks, Recreation and Historic Preservation; Department of Environmental Conservation, Department of Transportation, and local trail groups that have been instrumental in creating and maintaining trails along nearly a third of the canal system.

"I am pleased to be a partner in this important endeavor. I look forward to coordinating the planning, design and implementation of the trails program with local governments and trail organizations, "Mr. Tufo said. "I applaud the spirit of volunteerism and service by organizations such as the Erie Canal Trail Committee in the Mohawk Valley, E.R.I.E. (Erie's Restoration Interests Everyone) in Wayne County, the Erie Canal Parkway in Jordan, the Town of Camillus Erie Canal Park, and the Feeder Alliance in Glens Falls in laying the groundwork for the end-to-end trail."

WITH THE SHAKERS ON THE ERIE CANAL

By Elizabeth Hiddemen Rice, Perry, N.Y.

The First Erie Canal crossing the Genessee River on the West Side of Rochester. This Aqueduct is now a Highway Bridge.

Some of my research on the Utopian Communal Societies of the Shakers has been utilizing the microfilm copies of the material held at the Cleveland Western Reserve Historical Society. Among these papers one finds several journals kept, mostly by Shaker men of their travels on the varying Canals. This particular copy I am sending was written by Brother Seth Wells of his travel to Sodus, New York on the Erie Canal. The trip was made in 1832, 7 years after completion of the Erie at Buffalo, New York. The article is very descriptive and tells not only of the procedures on the Canal Boat "Statira," captained by Brooks, but other interesting facets, such as other passengers who boarded, not only with their families but with all their possessions. These people were emigrating to Illinois.

Seth Wells was an educated man who had joined the Shakers at New Lebanon, New York. The primary focus of this trip to Sodus at that time was for Seth to establish and set up the Shaker school system. Sodus Bay Community was located on Lake Ontario east of Rochester and had been established officially in 1826. There were many young people who had been attracted into the community and the need for schooling was felt.

E.H.R.

Bro. Seth Wells' Visit to Believers (Shakers) at Sodus, NY (near Rochester)

June 12, 1832 (Seth Wells)

Started from New Lebanon at 4 o'clock in the morning for Sodus and Port Bay. Proceeded on to GreenBush without any remarkable occurrence. Crossed the river in a Steam ferry Boat; and after a short stop in Albany, proceeded on to Schenectady where we arrived about 6 o'clock. p.m. We soon obtained a passage in the canal Boat Statira of Buffalo, *Capt. Brooks,* belonging to the Trader's Line. Started from Schenectady at quarter before 8 o'clock in the evening and went on a pretty good jag; all hands busy in arranging their baggage and other matters for the night. Some getting their suppers in the after cabin or dining room, some chatting on deck . . . some anxious about their baggage. Accordingly after seeing that everything belonging to us was in its proper place, and after they had swallowed their suppers in some way or other, the females retired to their berths in the rear of our cabin which was separated from us by a curtain across the cabin. After which the men retired to theirs. But there being more passengers than berths, some had to lie 2 in a berth. Brother Jethro and I retired to one about half past 10 but were unable to sleep in so cramped a situation. So after lying as still as two such old men could for an hour or more, br. Jethro got up, dressed himself and went on deck, leaving the berth wholly to me to rest, if I could, till the latter part of the night. After a while I got asleep and slept as far as I can learn about two hours, when we came into a lock. The noise of passing it waked me. We soon passed another lock, I then arose and gave up the berth to Jethro. I dressed myself and went on deck to write and finding all still above and below, I settled myself on deck in the bow of the boat by the side of a large Carboned lamp with this Journal and pencil in my hand and went to writing.

"Bull-Head" Canal Boat on the Enlarged Erie at Lock #50 in the 1890's. (Courtesy of the Canal Society of New York State.)

Aqueduct that Carried the Old Erie Canal Across Schoharie Creek at Fort Hunter.

Amsterdam

After passing Amsterdam, 16 miles from Schenectady, we were stopped in consequence of a heavy loaded boat being aground and several other boats near trying to pass by her, but could not. Here we were hindered till daylight, when the boat got off and we passed her and the others that had been stopped by her and passed on and crossed Schoharie Creek about half past 6 o'clock. A.M. 23 miles from Schenectady, having passed 9 locks from Schenectady. At half past 8 we saw Coughnsewaga on the North and passed Fulton on the south of the Mohawk. This is a new village rapidly building on the canal. At 10 we took breakfast and proceeded onto Spraker's Basin which we passed about 1/2 past 11., 36 miles from Schenectady. Proceeded on and arrived at Canajoharie 1/4 before 1:30 p.m. Went on to Fort Plains at 1/2 past 3. Past Openheim at 5; passed Danube, opposite Manheim at 7 o'clock. We then went on towards Little Falls and after passing 4 locks in succession with the most astonishing and very romantic views on each side, which for a night scene appeared gloomy and awfully sublime. We passed the Schoharie Aqueduct which crosses the river at Fort Hunter about 9 o'clock in the evening. After passing this wonderful aqueduct, we passed another lock which brought us thro' these falls, but the village being on the North of the Mohawk we could not have a fair view of it in the evening. We retired to bed about 11 o'clock

Mansion House, an Over-night stop for Erie Canallers at Rome, New York. The Black River Canal joined the Erie close by.

and Br. Jethro and I were glad enough to have some chance to sleep, and being now provided with each a berth we turned in and slept till about 4 o'clock, and then rose and came on deck, just in time to see the sun rise, having had nearly 5 hours sleep, a rare enjoyment on board a canal boat.

Utica

We found we had passed the German Flatts and Herkimer and were within about 3 miles of Utica, which we entered about 1/2 past 5, stopped about an hour and then passed on through the great city, which we left about 7 o'clock on Thursday morning, June 14th, 1832. Passed through whites' borough about 8. Came to Ortiskany at 9—Breakfasted at 10 - Arrived at Rome a little before 1 o'clock p.m., made a short stop and filled our water cask and went on - Arrived at New London about 3 p.m. This is the place on the confines of Wood creek near Oneida Lake, which has sprung up in the wilderness, by which it is still surrounded. Here are immense quantities of timber sawed into boards, plank and framing timber vc.vc. and transported down the canal to Albany, Troy and New York.

Oneida Creek

From thence we proceed on through Loomvis to Oneida creek, which we passed about 6 o'clock. Thence on through Lenox basin (114 miles from Schnectady.) Then on thro' to Canistota about 1/4 past 8. During the night we passed thro' New Boston, continuing to Kirkville, Manlius and Orville, (about 25 miles) and arrived at Syracuse 1st lock at the end of the long level of 69 1/2 miles about 4 o'clock on Friday morning, June 15. This lock is within 2 miles of Syracuse Village.

Syracuse

Here as we were going into the village I finished my letter which I had begun last night, read it to my companions, put in all our love, folded the letter and sealed it at the Village and slipped it into the P.O. This is a very flourishing village near the Salt works, where a vast deal of business seems to be carried on. We were detained at this place till 9 in the morning unloading some goods and taking on many more. We also took in 19 more passengers, men, women and children, consisting of 3 families, who were going on to the Country of Illinois 1700 miles distant. Many of their friends came on board with them to go on a few miles and take their last leave of them, so that we had about 30 additional passengers for a little distance. These emigrating passengers were respectable people and very civil, so that we were not much incommoded by them in the day time, but the night following (which was our last on the Canal) was far from being pleasant.

Salina Salt Works

After leaving Syracuse we had a very interesting view of the extensive salt works of Salina where salt is made by evaporation. The extended rows of double roof sheds for securing the salt from the weather, cover no less than 100 acres of ground. Between there long rows of sheds the spaces are covered with long rows of vats elevated from the ground and filled with salt water and exposed to the sun for evaporation, & so constructed as to slide under the sheds where they are secure from the rain and bad weather. The ground occupied by these sheds and vats is a large level plain, extending a great distance on each side of the Canal.

After leaving these salts works at Salina the course of the Canal has a northern direction which in a mile or two brought us to the village of Geddes. There we had a fair view of Salina on the south east, Onondaga Lake on the north and east, beyond which appeared the handsome village of Liverpool which together with the village of Geddes along the west of the Canal where we were now passing, afforded an extensive and beautiful prospect. Along the Canal in Geddes are an ex-

(Continued on Page 58)

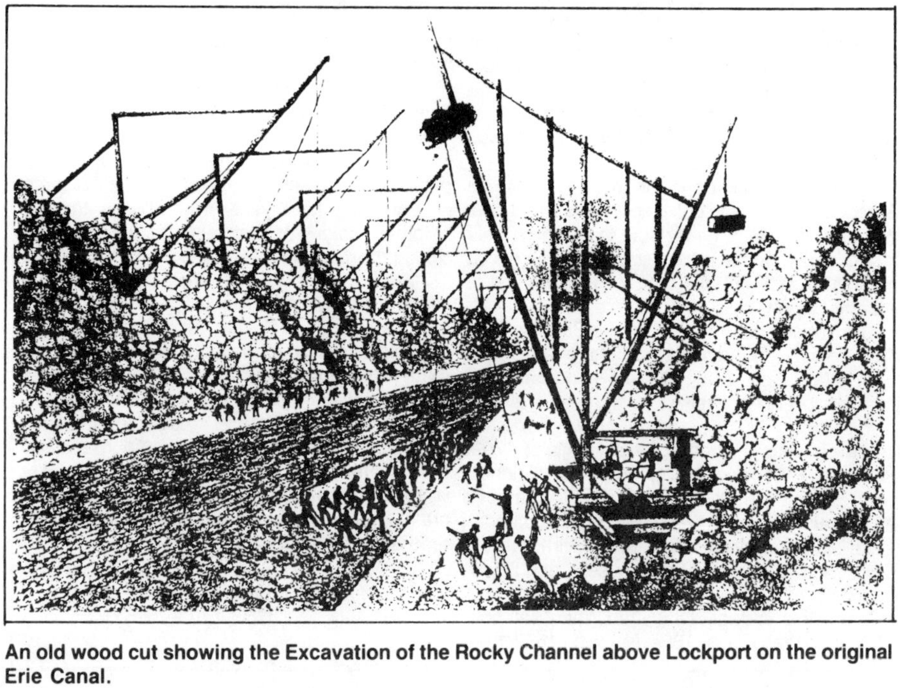

An old wood cut showing the Excavation of the Rocky Channel above Lockport on the original Erie Canal.

"THE SHAKERS"

The name "Shakers" was given to this sect (in derision) because of the bodily movements of the members in their worship and dancing ceremonies. An off-shoot of the Quakers in England, they were called officially "The United Society of Believers in Christ's Second Appearing."

Led by a strong-willed woman, "Mother" Ann Lee, the entire sect emigrated to America in 1774 and established themselves near New Lebanon in upper New York State. Their Motto: "Hands to Work and Hearts to God." They advocated celibacy and Christian communism, although they welcomed whole families into their membership.

The movement spread in upper New York State and New England and by the Mid-Nineteenth Century there were thousands of "Believers" in dozens of communal centers, including those mentioned in the accompanying article.

All worked diligently for the common good. The men were craftsmen who produced excellent furniture of a practical but distinctive form, the same in all the communes. Shaker furniture is in great demand today by antique collectors all over America.

During the past 150 years, the number of "Believers" has dwindled (no doubt because of their beliefs) to a mere handful. They have closed down most of their communes in upper New York State. Their commercial activities now consist of regular and widely advertised public "auctions" of Shaker furniture, which is being sold at fabulous prices!

Bill Shank

WITH THE SHAKERS ON THE ERIE CANAL

Getting a Fresh Horse out of the Stable on Board an Erie Freighter of the late Eighties. (Courtesy Canal Society of New York State.)

(Concluded from Page 57)

tensive row of buildings for boiling down the salt water by which means vast quantities of salt are made and barrelled up for market. Vast quantities of wood were piled along the Canal in this place for the use of their boilers.

Camillus

From this place we went on past Nine mile creek and Camillus, 150 miles from Schenectady. Thence we went on to Canton left Joran 6 o'clock p.m. 35 miles from Lyons, went on to 5 miles-Jordan 6-miles-Weed'sport 6 miles-Center Port-1 mile-Port Byron 2 miles- Thence went on to Lake Port 6 miles- Thence on thro' Montezuma and the Cayuga Marshes. All these places from Jordan to Clyde we passed in the night. The weather very warm, which with our additional number of passengers who came on board at Syracuse rendered our cabin no very comfortable place to sleep in- I undertook to find a place in the midships where the goods were stowed away. There being a number of beds belonging to the families moving to Illinois, which were tied up in bundles, also some bags vc. We tumbled them together and myself and two of the boatmen, tumbled down near one of the side doors of the boat, leaving both of the side doors open, so as to have a circulation of air-there being no windows in the midships. We had lain but a very short time when we came into a lock-it being a descending lock which we had to pass after leaving Port Byron where we had been detained till near midnight in unloading goods. As the water was let out of the lock, and the boat settled down into it, there was a leak in the side of the lock from which a stream of water issued directly into the side door near where we lay - The boatmen scrambled up with some difficulty shoved the boat along so as to prevent the water from coming in at the door, but could not shut the door till we got clear of the lock, when all was made tight lest we should get another spattering at the next lock. But I soon quit the midships and went on deck. There I found a number of passengers who had left the cabin, being no better able to stay there than I was to stay in the midships.

Mosquitoes

We were now passing through the Cayuga Marshes, and were greatly annoyed with mosquitoes, both in the cabin and on deck. The passengers are on deck, fighting and killing the troublesome bites made many diverting speeches about them, one was for having them harnessed in and make them help draw the boat - Another proposed to leave them boxed up and sent off as a present to General Jackson - another not quite so good natured as his companion setting his foot on a grindstone which lay on deck wished they had that grindstone hung to their necks, he would sink them where they would never rise again. Thus we passed thro' these marshes, deprived of sleep, annoyed by mosquitoes and sweating with the heat of the weather.

We find the Shakers

We arrived at Clyde at 1/2 past 5 on Saturday morning - 9 miles from Syracuse. The boat having stopped at Galen, about 4 1/2 miles below Lyons to take in 2 women passengers. We discovered a wagon standing in the street at a distance, with Believers (Shakers) in it — they had discovered us and were looking at us. I immediately jumped off the Boat which was then on the opposite side of the Canal and ready to start on. I requested the Captain to wait a few minutes till I could go and speak to them. I crossed the bridge near by and soon met Daniel Dryer, and we walked up to the wagon and found Lucy Brown, Tina Seaton and a young Sister with them, who had come out on some business the day before and stayed at Smith Tindal's whose wife is a believer (Shaker). They had not received our letter, but had been daily looking for it for a week past, and wondered why it did not come. I returned to the Boat and we went on to Lyons. They rode along little way and called at the house of an acquaintance, wrote a letter to Elder J. and Eldress E. and sent a messenger in haste, and then proceeded on to Lyons and arrived about the same time we did, met us at the lock and they took us and our baggage to Benton's tavern, where we arrived soon after 9 o'clock. We all took breakfast together, and tarried and rested ourselves till about 3 o'clock when Lucius Southwick arrived from Sodus, when after bating (baiting"?) his horses we all started together for Sodus. D. Dryer took Sister Olive, Susanna and Lucy B. in his wagon and Lucius took Br. Jethro and me, and the 2 sisters, Tina and Lydia Dole.

MORE ON THE SHAKERS

Perhaps I can contribute some information that will be more accurate, as I am a long-time collector, researcher, scholar on the subject of the Shakers, having been associated with the restoration of the Shaker community, Pleasant Hill, Ky., and on the Advisory Council there.

When the small band of Shakers came to America in 1774, "the entire sect" was only 9 people, and they were finally able to establish their first group together near Albany, New York at Niskayuna in 1776, later known as Watervliet in Albany County.

The various communities of the Shakers, and their dates of founding, are a total of 24. In the East: Watervliet—1787, in New York State and then Mount Lebanon (also formerly known as New Lebanon) 1787, also in New York. In Massachusetts: Hancock (near Pittsfield) 1790, Harvard, 1791, and Tyringham, 1792. Connecticut: Enfield, 1790. New Hampshire: Canterbury, 1792, and Enfield, 1793. Maine: Sabbathday Lake, 1794, and Gorham, 1808, Alfred, 1793. Shirley in 1793, also in Massachusetts, and Savoy, 1817.

Then Union Village, Ohio, and Watervliet, both 1806, and Whitewater, 1824, N. Union 1822. Kentucky: Pleasant Hill, 1806, and South Union, 1807. Indiana: West Union, 1810.

Later: in New York, Sodus Bay, 1826, and Groveland, 1836. In Narcoossee, Florida, 1896, and White Oak, Georgia, 1898.

Due to their dwindling numbers, today at Sabbathday Lake, Maine are the few left, also some new converts have joined, but officially the last "Mother" colony or spiritual head at Canterbury closed the rolls some years back.

So your statement that "their commercial activities now consist of regular and widely advertised "auctions" of Shaker furniture, which is being sold at fabulous prices!" - is misleading as these restorations and the existing Shakers at Sabbathday Lake *do not hold such auctions.* Auctions are held by dealers, often at sites as New Lebanon has been one such site. But such auctions are also held in New York City, elsewhere. The Shaker villages/sites are not selling off their furniture but are trying to collect and maintain sites-please do read about Hancock!

Susan Jackson Keig, FSTA
860 Lake Shore Drive
Chicago, IL 60611

We arrived at Sodus about 1/4 before 8 o'clock, and were joyfully received and made welcome.

Sodus

Sabbath morning June 17th after breakfast we met the Elders together and gave them the love, which they rec'd with evident feelings of thankfulness conversed with them for some time. As the time of their public meeting drew nigh the world began to collect in considerable numbers — We did not attend the public meeting, but being in the upper part of the meeting house we had a view of their exercises. We found they had a very good gift in singing, laboring in the step manner and marching. During an intermission of their exercises Joseph Pelham addressed the world, had a very good gift . . At 3 o'clock p.m. the families collected and had a meeting without the world. We attended. Elder Jeremiah opened the meeting, expressed his thankfulness upon the occasion of meeting with us who were from the Church at New Lebanon - the Mother Church and center of union. He endeavored to impress the minds of the Believers with a sense of the necessity of feeling their faith joined to the center of union where the gift of their protection and blessing centered and from when all their blessings and their spiritual strength and protection proceeded . . He observed to them that we had brought a treasure of love from the Church and Society at New Lebanon and hoped they would be able to receive it as a rich treasure from the fountains at New Lebanon. He appeared much affected with a sense of thankfulness for our visit which they considered as a great privilege and spoke of this need they all had of the love, union and blessing of the Church at New Lebanon as the fountain to which they looked for strength and protection.

They then sang the following welcome song, which had been prepared that morning by Jesse Leonard for the occasion, with much feeling.

In humiliation we ask while we sing,
Does Zion remember her little offspring?
Or love from the fountain have you got today.
To give to your children who live at Port Bay.

Our vessels are open, we want the good fruit
The twig of the vine receives sap from the root
We will be good children and some future day,
May gather good fruit from its bud in Port Bay

We pray for your blessing, and may you be blessed
With health, frean and comfort we wish you the best,
Those heavenly pillars we truly can say
We thankfully welcome you all to Port Bay.

We then delivered the love and felt an uncommon gift in expressing the great interest felt in the Church and Society at New Lebanon and for their blessing and prosperity in the gospel which had a great effect upon their feeling, and caused weeping throughout the assembly. After we had all spoken our feelings, Elder Jeremiah expressed his thankfulness with much feeling, and told them they might kneel down and cry as much as they were a mind to. Accordingly they all fell on their knees and wept and many cried aloud. After rising and speaking with much feeling and solemnity, Elder Jeremiah told them they might rejoice and such leaping and rejoicing we have rarely ever seen anywhere - Singing and dancing and shaking and rejoicing and hugging and kissing followed - then marching vc.

After meeting we met the Elders and had a good meeting and supported union with our brethren and sisters in the church at New Lebanon with great satisfaction. This had been agreed upon before we left New Lebanon. (End)

STEAM TUG TRAPPED

Ex Philadelphia and Reading Railroad steam tug "Catawissa" sits on the bottom between Locks 3 and 4 of the Waterford Flight in Waterford, NY. She was built by Harlan & Hollingsworth at Wilmington, DE in 1896 and was one of nine sister tugs used in the Philadelphia to New England coastal coal trade. The boiler and 1000 horsepower triple expansion engines of this 637 ton 159′ by 29′ vessel remain in place. After World War II she was relegated to tank cleaning duties first by the Bethlehem Hoboken Shipyard as "Beth Tank Ship No. 2" and was finally working for Standard Tank Cleaning Co. as "Tank Master No. 1." In the fall of 1992 "Catawissa" was sold to midwest interests to be converted into a restaurant, but was found to have too great a draft to negotiate the Erie Canal. Her future is uncertain at this time. Bill McKelvey reports that the "Catawissa" is still there (1994) and her future status is still unknown. (Photo courtesy of Paul Spunable, New York State Thruway Authority.)

NEW YORK CANALS BEFORE THE ERIE

Although 1825, the year of completion of the Erie Canal, is often cited as the beginning of the canal era in the United States, navigation between the Hudson River and the Great Lakes goes back to prehistoric times, and canalization projects in the Erie Canal corridor are as old as the Constitution. 1992 will be the bicentennial year of the Western Inland Lock Navigation Company, which provided navigational improvements between Albany and Oswego, on Lake Ontario, from 1792 until it was superseded by the Erie. Planning for the bicentennial celebration is being coordinated by the New York State Museum. Canal societies, communities, companies, and individuals interested in participating should make contact with: Philip Lord, Jr., Room CEC 3097, New York State Museum, Empire State Plaza, Albany, New York 12230. Early works like those shown in the above artist's drawing opened westward navigation in the 1790s. (From the Confederation Life Collection, reproduced by permission.)

OTHER PUBLICATIONS OF THE AMERICAN CANAL AND TRANSPORTATION CENTER

THE AMAZING PENNSYLVANIA CANALS – 160th Anniversary Edition – By William H. Shank (1991). A much expanded variation of many previous printings. 125 illustrations, and tables of locks and mileages on most of the principal canals in the State, never previously gathered together in one volume. Four-color cover; two-color interior; 128 pages; a definitive work.

TOWPATH GUIDE TO THE CHESAPEAKE AND OHIO CANAL – By Thomas F. Hahn, (1994). A fully illustrated, historical commentary and mile-by-mile directory for the entire 184-mile length of the C. & O. Canal Towpath from Washington, D.C. to Cumberland, Maryland. Excellent maps included. 226 pages.

HISTORIC BRIDGES OF PENNSYLVANIA – By William H. Shank, (1986 Edition). Traces the development of the bridge-building arts from the time of the first covered bridge in America, to modern bridges of the 20th Century. Biographies of such famous bridge builders as John Roebling, Theodore Burr, Charles Ellet and Ralph Modjeski included. Profusely illustrated.

PENNSYLVANIA TRANSPORTATION HISTORY – A Supplement – By William H. Shank (1990). In this book, Mr. Shank has discussed media and devices not covered in his other books. Early river craft, rope ferries, steam boats, inclined planes, gravity railroads, early steam locomotives, horse cars, cable cars, trolley cars, elevated rail and subway systems, and air-travel devices are included. An 8-1/2" x 11" book with 72 pages and approximately 100 old photos, drawings and tables. Four-color cover.

GREAT FLOODS OF PENNSYLVANIA–A TWO-HUNDRED YEAR HISTORY – W. H. Shank, (Seventh Printing, 1993). Data, photos and non-technical text on all major floods in the Keystone State since records have been kept. Full chapters on the Johnstown Flood and the disastrous floods of 1936, 1955 and 1972 are included. A definitive work.

INDIAN TRAILS TO SUPERHIGHWAYS – By William H. Shank, (1988 Edition). History of the development of Pennsylvania's historic roads and the many interesting vehicles used on them. Descriptions of Braddock's Road, Forbes' Road, National Highway, Lancaster Turnpike, Plank Roads, Corduroy Roads, William Penn Highway, Lincoln Highway, Pennsylvania Turnpike and Keystone Shortway. Recent PennDOT plans for completion of Pennsylvania's Interstate Highway System.

THE BEST FROM AMERICAN CANALS – Number 2 (1980-1983). A publication of the American Canal Society (1990) reprinting articles on historic and currently operating canals in USA and overseas. Featured are the Panama Canal, the Trent-Severn Waterway, and canals in Europe, Thailand and China. 140 illustrations and maps included, and a worldwide index. An 8-1/2 x 11 paperback with 88 pages.

THE BEST FROM AMERICAN CANALS – Number 3 (1983-1986). A publication of the American Canal Society (1986), featuring the Canals of New England, metropolitan New York area, Erie Canal, the I. & M. Heritage Corridor, the Tenn-Tom, St. Lawrence Seaway, and Canals of Sweden, Germany and India. 145 illustrations and maps, and a world-wide index. An 8-1/2 x 11 paperback with 88 pages.

THE BEST FROM AMERICAN CANALS – Number 4 (1986-1988). A publication of the American Canal Society (1989), featuring maps of the principal canals of the USA, Erie Canal, Inclined Planes, Morris Canal, S. & T. Canal, Shinnecock Canal, Champlain Canal, Navigation on the Mississippi, Kanawha, Kentucky , Fox, and Muskingum Rivers. 8-1/2 x 11 paperback, 88 pages.

THE BEST FROM AMERICAN CANALS – Number 5 (1989-1991). A publication of the American Canal Society (1991) including a complete history of the Panama Canal, up-date on the Rhine-Main-Danube Canal, description of several Chinese canals, and navigation tips on the Black River, the Hennepin, the Alabama and the Florida waterways. 8-1/2" x 11" paperback, 88 pages.

THE BEST FROM AMERICAN CANALS – Number 6 (1991-1993). A publication of the American Canal Society (1993) featuring many canals in U.S.A.; international canal meets; cruises on the Mississippi, Lake Michigan, Louisiana, Canada, Japan, Portugal, Ireland and the Corinth Canal.

HISTORY OF THE YORK-PULLMAN AUTOMOBILE, 1903-1917 – By William H. Shank, (1970). History of the "Six-Wheeler" Pullman, and its successors, which almost made York, Pa. the automotive capital of the United States. History of the early automotive industry in Eastern Pennsylvania also included. Profusely illustrated. An ideal gift for antique car buffs.

WHEN HORSES PULLED BOATS – By Alvin F. Harlow. Third reprinting, 1994, of Harlow's little-known canal book, written for school students in 1936. An excellent exposition of the historic canal era in the USA. Introduction by William H. Shank. Canal bibliography included. Well illustrated with sketches by Orson Lowell and Philip Hoffmann.

VANDERBILT'S FOLLY–A HISTORY OF THE PENNSYLVANIA TURNPIKE – W. H. Shank, (Tenth Printing, 1993). The railroad war of 1880-85 which created the tunnels and roadbed for the present turnpike. History of the Turnpike, 1940-1993, included.

THE COLUMBIA-PHILADELPHIA RAILROAD AND ITS SUCCESSOR – William Hasell Wilson, 1896. This book is an on-the-spot account of the building of one of the oldest railroads in America by its chief engineer, later resident engineer for the Pennsylvania Railroad, who purchased it from the Pennsylvania Canal Commissioners. This 1992 reprint is fully illustrated with 45 old photos, maps and drawings. (Second printing).

THE CANALLER'S SONG BOOK – By William Hullfish, Music Instructor, Writer, Singer (Third Edition, 1993). This 88 page, 8-1/2 x 11 book contains forty historic canal ballads (words and music) collected from old records in the northeastern USA over a ten year research period. Illustrated with old canal drawings. Said to be the most comprehensive collection of canal songs ever published.

PICTURE-JOURNEY ALONG THE PENNSYLVANIA MAIN LINE CANAL, 1826-1857, By Philip J. Hoffmann, P.E. (1993). Edited by William H. Shank, P.E. Full-color drawings of entire state-owned route, Philadelphia to Pittsburgh. 8-1/2" x 11" paperback. 80 pages. Full Hoffmann biography included.

TOWPATHS TO TUGBOATS – A History of American Canal Engineering. By Shank, Mayo, Hahn and Hobbs (Fifth Printing, 1995). The works of such famous Canal Engineers as Benjamin Wright, Canvass White, Charles Ellet, William Hamilton Merrit, George Washington Goethals are detailed – with the canals they built. The Erie, the Welland, the "Soo," the Panama, the St. Lawrence Seaway and the Tenn-Tom are among the many waterways described in detail. A 72-page, 8-1/2" x 11" book, the publication contains more than 130 drawings, maps and photographs in USA, Canada and overseas.

Inquiries may be directed to the American Canal and Transportation Center, 809 Rathton Road, York, PA 17403. Price list and discount schedule available.